FMGEMS

Clinical Sciences

Examination Review

FMGEMS

Clinical Sciences
Examination Review
Fourth Edition

700
Questions
and
Answers

MEPC
MEDICAL EXAMINATION
PUBLISHING COMPANY

Medical Examination Publishing Company
A Division of Elsevier Science Publishing Co., Inc.
655 Avenue of the Americas,
New York, New York 10010

This book is printed on acid-free paper.

ISBN 0–444–01561–2

Current printing (last digit):
10 9 8 7 6 5 4 3 2 1

Manufactured in the United States of America

Contents

Preface

FMGEMS Basic Sciences Examination Review and *FMGEMS Clinical Sciences Examination Review* are designed as study aids for candidates preparing to take the Foreign Medical Graduate Examination in the Medical Sciences (FMGEMS).

The clinical sciences portion of the examination consists of a battery of objective-type questions in a variety of formats, related to the following subject areas: internal medicine, obstetrics and gynecology, pediatrics, preventive medicine and public health, psychiatry, and surgery. The questions in this book have been selected in accordance with the content outline furnished by the ECFMG to cover the full range of clinical scientific knowledge.

The format of this book is similar to that of the FMGEMS examination, with multiple-choice questions, matching problems, and case histories with related questions within each discipline. All questions are referenced to current editions of major textbooks within each field, and all are keyed to detailed explanatory answers.

The companion volume, *Basic Sciences,* is to be used to prepare for the basic sciences portion of the examination. *Basic Sciences* and *Clinical Sciences* are designed to be used together. Their use, in conjunction with textbooks and other educational materials, will certainly help the prospective examinee by giving him or her a clearer idea of the depth and breadth of his or her knowledge and retention. Additionally, the reader will gain a considerable psychological edge because the style and format of the tests will be familiar and the material will be accessible and easily recalled.

Contributors

Michael A. Baker, MD, FRCP(C)
Professor of Medicine, University of Toronto Faculty of Medicine; Head, Department of Haematology, Director, Oncology Programme, Toronto General Hospital, Toronto, Ontario, Canada

Carlyle H. Chan, MD
Associate Professor and Director of Residency Education, Department of Psychiatry and Mental Health Sciences, Medical College of Wisconsin, Milwaukee, Wisconsin

Theodore P. Haddox, Jr, MD
Assistant Professor, Department of Obstetrics and Gynecology, Marshall University School of Medicine, Huntington, West Virginia

Faculty Members with **Richard H. Hart, MD**, Chairman Department of Public Health and Preventive Medicine, Loma Linda University School of Medicine, Loma Linda, California

Victor LaCerva, MD
Clinical Assistant Professor, Department of Pediatrics, University of New Mexico Medical Center, Albuquerque; MCH Medical Director, Maternal and Child Health, State of New Mexico Public Health Division, Santa Fe, New Mexico

Sherwood C. Lynn, Jr, MD
Associate Professor, Department of Obstetrics and Gynecology, Marshall University School of Medicine, Huntington, West Virginia

Michael H. Metzler, MD
Associate Professor, Departments of Surgery and Anesthesiology, Chief, Division of General Surgery, University of Missouri Health Sciences Center, Columbia, Missouri

Harry Prosen, MD
Professor and Chairman, Department of Psychiatry and Mental Health Sciences, Medical College of Wisconsin, Milwaukee, Wisconsin

FMGEMS
Clinical Sciences
Examination Review

1 Medicine

Michael A. Baker, MD, FRCP(C)

DIRECTIONS (Questions 1–47): Each of the questions or incomplete statements below is followed by five suggested answers or completions. Select the **one** that is best in each case.

1. Evaluation of calcific mitral stenosis by M-mode echocardiography is likely to show
   A. increased diastolic slope
   B. absence of valve closure in mid-diastole
   C. closure of valve in late diastole
   D. presence of normal M-shaped configuration
   E. absence of valve closure in systole

2. The normal mean pulmonary artery wedge pressure at rest is between
   A. 1 and 10
   B. 1 and 25
   C. 10 and 20
   D. 15 and 25
   E. 0 and 2

3. An asymptomatic 45-year-old man has a short, soft systolic murmur, loudest at the left upper sternal border, associated with an ejection click that has been documented since childhood. The most likely explanation is
    A. patent ductus arteriosus
    B. coarctation of the aorta
    C. ventricular septal defect
    D. Epstein's anomaly of the tricuspid valve
    E. pulmonic stenosis

4. A 70-year-old woman with a Starr-Edwards prosthetic aortic valve develops fever and a new heart murmur. Laboratory findings are likely to include all of the following **EXCEPT**
    A. elevated erythrocyte sedimentation rate
    B. positive blood cultures
    C. erythrocytosis
    D. leukocytosis
    E. proteinuria

5. A 51-year-old renal transplant patient is admitted with possible transplant rejection. The cardiogram (lead II) is shown in Figure 1.1. What is the most likely cause of the changes seen?
    A. Hyperkalemia
    B. Anterior wall infarction
    C. Second-degree heart block
    D. Hypercalcemia
    E. Acute pericarditis

Figure 1.1

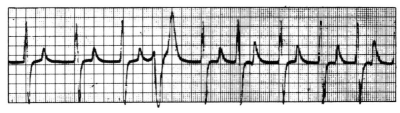

6. A 65-year-old man is well except for recent onset of angina. It persists on exertion in spite of intermittent use of nitroglycerine. The next step in management for this patient is likely to be
   A. percutaneous transluminal coronary angioplasty
   B. coronary artery bypass surgery
   C. a beta-blocker and calcium antagonist
   D. laser angioplasty
   E. controlled-rate pacemaker insertion

7. A 70-year-old man develops fever and pleuritic anterior chest pain three weeks after an uneventful anterior wall myocardial infarction. There are no new Q waves and there is no rise in CK-MB enzymes. The most likely diagnosis is
   A. reinfarction
   B. Dressler syndrome
   C. lobar pneumonia
   D. cardiac tamponade
   E. viral pericarditis

8. A 52-year-old patient develops a cerebrovascular accident three months after a myocardial infarction. The chest x-ray is shown in Figure 1.2. What is the most likely diagnosis?
   A. Pericardial effusion
   B. Mitral stenosis
   C. Ventricular aneurysm
   D. Subacute bacterial endocarditis
   E. Papillary muscle rupture

9. A 25-year-old intravenous drug abuser presents with a four-week history of dry cough and dyspnoea. There is a diffuse interstitial pattern on x-ray. The most likely cause is
   A. pulmonary emboli
   B. *Pneumocystis carinii*
   C. Kaposi's sarcoma
   D. Legionnaire's disease
   E. Lyme disease

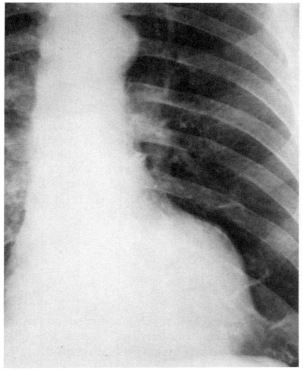

Figure 1.2

10. A 60-year-old man has transient ischemic attacks. Progression to stroke may be best prevented by
   A. lowering blood pressure
   B. oral coumarin anticoagulants
   C. conversion of any arrhythmias to sinus rhythm
   D. oral aspiration
   E. intracranial carotid endarterectomy

11. A 21-year-old man had a history of diabetes insipidus and
progressive shortness of breath. He had a lytic lesion in the
distal femur. Part of his chest x-ray is shown in Figure 1.3.
What is the most likely diagnosis?
   A. Eosinophilic granuloma
   B. Tuberculosis
   C. Bronchogenic carcinoma
   D. Viral pneumonia
   E. Multiple myeloma with pneumonia

Figure 1.3

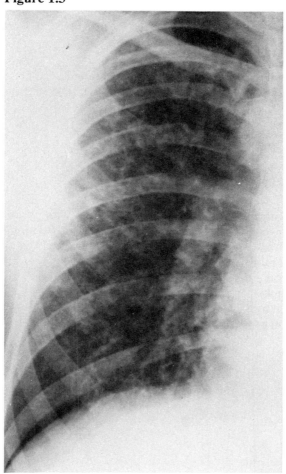

12. Ocular abnormalities associated with the Wernicke-Korsakoff syndrome include
    A. papilloedema
    B. Horner's syndrome
    C. cotton wool exudates
    D. horizontal and vertical nystagmus
    E. proptosis

13. L-dopa treatment of Parkinson's disease is most often associated with dramatic improvements in
    A. mood
    B. tremor
    C. long-term disease progression
    D. strength
    E. hypokinesia

14. A 32-year-old pregnant woman presents with unilateral nerve deafness, hyperpigmented skin patches, and multiple subcutaneous tumors. The most likely diagnosis is
    A. malignant melanoma with metastases
    B. amniotic fluid emboli
    C. neurofibromatosis
    D. Osler-Weber-Rendu disease
    E. Sturge-Weber syndrome

15. A patient with myasthenia gravis being treated with pyridostigmine develops increased weakness, nausea, vomiting, sweating, and bradycardia. The best management includes
    A. thymectomy
    B. corticosteroids
    C. plasmapheresis
    D. reduced pyridostigmine
    E. addition of neostigmine

16. Chronic obstructive pulmonary disease from α-1-antitrypsin deficiency is associated with
    A. MM phenotype
    B. ZZ phenotype
    C. X-linked recessive inheritance
    D. no available treatment
    E. no other organ involvement

17. The anemia of chronic renal failure is typically
    A. responsive to erythropoietin
    B. macrocytic
    C. secondary to iron loss
    D. transient
    E. responsive to DDAVP

18. Which of the following leads to a better graft survival following kidney transplantation?
    A. ABO incompatibility
    B. Cadaveric versus living donors
    C. Presensitization to donor tissue by skin graft
    D. Substitution of cyclophosphamide for azathioprine for graft rejection
    E. Blood transfusions for the recipient

19. A 16-year-old black female is admitted with a history of chronic leg ulcers and intermittent severe pain in the arms and legs. Investigation includes the x-ray shown in Figure 1.4. What is the most likely diagnosis?
    A. Traumatic fracture
    B. Hereditary β chain defect
    C. Hereditary α chain defect
    D. Ankylosing spondylitis
    E. Plasmacytoma

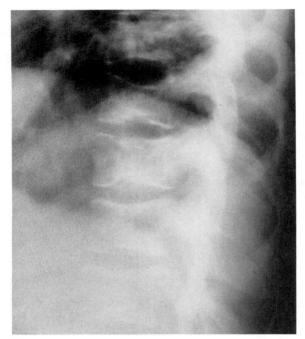

Figure 1.4

20. Intravenous amphetamine abuse is most likely to cause renal disease associated with
   A. membranous glomerulonephritis
   B. proliferative glomerulonephritis
   C. necrotizing vasculitis
   D. acute tubular necrosis
   E. nephrotic syndrome

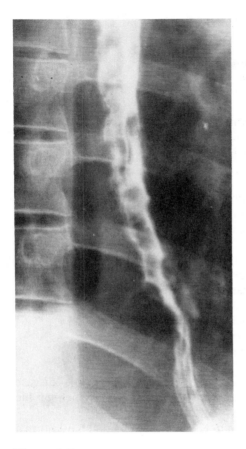

Figure 1.5

21. A 42-year-old musician is admitted with hematemesis and, on examination, has hepatomegaly. Part of his upper GI barium swallow study is shown in Figure 1.5. What is the diagnosis?
    A. Bronchogenic carcinoma
    B. Esophageal stricture
    C. Reflux esophagitis
    D. Esophageal carcinoma
    E. Esophageal varices

22. In acute, uncomplicated cystitis, 90% of infections are due to
   A. *Proteus* organisms
   B. pseudomonas
   C. *Klebsiella* organisms
   D. *Escherichia coli*
   E. *Serratia* organisms

23. A 51-year-old man with gross hematuria has erythrocytosis, enlarged kidneys, and a family history of renal disease. The most likely diagnosis is
   A. polycystic kidneys
   B. hypernephroma
   C. malignant hypertension
   D. bilateral ureteral calculi
   E. medullary cystic disease

24. Which of the following patients would be the best candidate for an allogeneic bone marrow transplant?
   A. Acute lymphoblastic leukemia, aged 30
   B. Acute myeloblastic leukemia, aged 30
   C. Chronic myelogenous leukemia, aged 20
   D. Chronic lymphocytic leukemia, aged 45
   E. Myelodysplastic syndrome, aged 45

25. A 20-year-old woman with stage IV Hodgkin's disease receives chemotherapy with MOPP-ABVD. Which of the following statements is true?
   A. She will be unable to become pregnant and have normal children
   B. She has more than a 100-fold chance of contracting leukemia compared to the general population
   C. Her intellectual function will be impaired
   D. She has about a 70% chance of relapsing with Hodgkin's disease within five years
   E. She should receive prophylactic radiation to the central nervous system

26. A 40-year-old woman with mild Von Willebrand disease requires minor surgery for a skin lesion. Optional management would include
   A. intravenous stored plasma during surgery
   B. cryoprecipitate every 12 hours for one day prior to and two days post-surgery
   C. 100 units of factor VIII during surgery
   D. factor IX complex given topically postoperatively
   E. Deamino-D-arginine vasopression intravenously just before surgery

27. Which paraneoplastic syndrome suggests the presence of epidermoid (squamous cell) carcinoma of the lung?
   A. Inappropriate secretion of antidiuretic hormone
   B. Ectopic secretion of ACTH
   C. Ectopic parathyroid hormone
   D. Myasthenic syndrome
   E. Periostitis

28. In adults, the most frequent diagnosis of primary tumors of the anterior mediastinum is
   A. germ cell neoplasms
   B. lymphomas
   C. endocrine tumors (thyroid and parathyroid)
   D. thymic lesions (cysts, hyperplasia, thymoma)
   E. mesenchymal tumors

29. Carcinoma of the stomach is most likely to metastasize to
   A. liver
   B. peritoneum
   C. lungs
   D. adrenal
   E. bone

30. In investigating a patient for possible hepatocellular carcinoma, the best tumor marker would be
   A. human chorionic gonadotrophin
   B. α-fetoprotein
   C. carcinoembryonic antigen
   D. alkaline phosphatase
   E. S-100 antigen

31. Methotrexate exerts its cytotoxic effects through
   A. alkylating DNA strands
   B. inhibition of dihydrofolate reductase
   C. acting as a false base analog
   D. interference with microtubular assembly
   E. binding to reverse transcriptase

32. A postmenopausal woman has a primary breast cancer resected, but lymph node dissection of the axilla reveals several nodes that are positive for metastatic disease. Additional therapy should include
   A. tamoxifen
   B. ovarian ablation
   C. radical mastectomy
   D. adjuvant radiation to the axilla
   E. cyclophosphamide, methotrexate, and fluorouracil

33. A young woman with acute uveitis and erythema nodosum is found to have bilateral symmetric hilar adenopathy and paratracheal adenopathy without chest symptoms. The most likely diagnosis is
   A. tuberculosis
   B. AIDS
   C. Lyme disease
   D. sarcoidosis
   E. rheumatoid arthritis

34. Intensive care unit patients with poor nutrition and receiving broad spectrum antibiotics are likely to exhibit
   A. increased factor VII activity
   B. increased vitamin K excretion
   C. shortened bleeding time
   D. reduced synthesis of factor VIII
   E. reduced synthesis of prothrombin

35. A 45-year-old man with arthralgias, diarrhea with malabsorption, increased skin pigmentation, and a seventh nerve palsy has a clinical remission after one year of treatment with trimethoprim-sulfamethoxazole. His diagnosis is
   A. intestinal lymphoma
   B. miliary tuberculosis
   C. Whipple's disease
   D. tropical sprue
   E. ulcerative colitis

36. The mechanism of action of sucralfate in treatment of peptic ulcers is thought to be
   A. neutralization of hydrochloric acid
   B. inhibition of basal acid secretion
   C. antagonism of the H-2 receptor
   D. binding to granulation tissue
   E. stimulation of prostaglandin E

37. A 25-year-old woman with mild persistent unconjugated hyperbilirubinemia is most likely to have
   A. Crigler-Najjar syndrome
   B. Dubin-Johnson syndrome
   C. hereditary glucuronyl transferase deficiency
   D. acute hemolytic anemia
   E. chronic active hepatitis

38. A 65-year-old man has a history of progressive fatigue over several years and increasing bone pain. He has firm splenomegaly and pancytopenia. Bone marrow aspiration shows infiltration with large cells with fibrillar cytoplasm. The most likely diagnosis is
   A. malignant lymphoma
   B. Gaucher's disease
   C. chronic myelogenous leukemia
   D. Niemann-Pick disease
   E. phenylketonuria

39. Treatment of hyperlipidemia with a fibric acid derivative such as clofibrate is indicated for
   A. young women with a family history of heart disease
   B. any men with very high HDL cholesterol
   C. angina patients with high postprandial triglycerides
   D. men with low levels of apoprotein B
   E. middle-aged men with very high LDL cholesterol

40. A 40-year-old man with a history of ethanol abuse develops blistering of light-exposed skin and brown urine. This condition is best managed by
   A. intermittent phlebotomy
   B. institutionalization
   C. daily colchicine
   D. oral and topical corticosteroids
   E. cholecystectomy

41. Diabetes insipidus syndromes are characterized by
   A. urine specific gravity above 1.010
   B. urine osmolality over 300 mosm/kg
   C. contraction of the bladder
   D. elevated serum sodium concentration
   E. decreased glomerular filtration rate

42. Which of the following features is most characteristic of non-insulin-dependent, type II diabetes?
   A. Obese body weight
   B. Age of onset usually under 30
   C. Ketosis is common
   D. Circulating islet cell antibodies
   E. Association with autoimmune phenomena

43. Osteomalacia in chronic renal failure is associated with
   A. high serum calcium
   B. low serum phosphorus
   C. high serum alkaline phosphatase
   D. low serum parathyroid hormone
   E. hypothyroidism

44. Nonketotic hyperosmolar syndrome is best characterized by
   A. hypervolemia
   B. sudden onset
   C. increased renal glucose excretion
   D. normal serum sodium and potassium
   E. ketonemia

45. A 60-year-old woman has a three-month continuous history of morning stiffness, and pain and swelling of the second and third metacarpalphalangeal joints of both hands. The best diagnosis is
   A. definite rheumatoid arthritis
   B. possible scleroderma
   C. probable systemic lupus erythematosis
   D. definite osteoarthritis
   E. classic psoriatic arthritis

46. The single most useful diagnostic laboratory test for systemic lupus erythematosis is
   A. the fluorescent antinuclear antibody test
   B. detection of LE cells
   C. measurement of C3 complement component
   D. antibodies to native DNA
   E. the lupus anticoagulant

47. Photochemotherapy utilizing psoralens is best reserved for
    A. exfoliative dermatitis
    B. primary malignant melanoma
    C. systemic sclerosis
    D. severe psoriasis
    E. severe eczematous dermatophytosis

DIRECTIONS (Questions 48–71): This section consists of clinical situations, each followed by a series of questions. Study each situation and select the **one** best answer to each question following it.

Questions 48–51: A 42-year-old African American woman complains of mild headaches for three months, relieved easily by mild analgesics. A careful history reveals no associated symptoms. She has had three pregnancies which were uneventful except for a requirement for diuretics in the last three months of her last pregnancy. There have been no past illnesses of significance except for cystitis requiring antibiotics on one occasion several years ago. Her father and a paternal uncle died of "heart attacks" at ages 43 and 46, respectively.

Upon physical examination, the patient is a slim woman with a BP of 180/120 in both arms after quiet resting. There was mild venous "nicking" on fundal examination. Examination of the heart was unremarkable, as was examination of all other systems.

Initial laboratory examination revealed trace proteinuria, a normal cardiogram, and a normal chest x-ray.

48. The most likely causes of this woman's hypertension are
    A. pheochromocytoma or renal artery disease
    B. renal artery disease or brain tumor
    C. brain tumor or Cushing's syndrome
    D. Cushing's syndrome or essential hypertension
    E. essential hypertension or chronic pyelonephritis

49. The following diagnostic tests would be indicated **EXCEPT**
   A. intravenous pyelogram
   B. bone marrow aspiration
   C. creatinine clearance
   D. plasma renin activity
   E. electroencephalogram

50. If repeated testing were to reveal hypokalemia, the differential diagnosis would then include all of the following **EXCEPT**
   A. pheochromocytoma
   B. primary aldosteronism
   C. Cushing's syndrome
   D. renal parenchymal disease
   E. diuretic abuse

51. If the patient is shown to have essential hypertension, a reasonable first treatment regimen might include all of the following **EXCEPT**
   A. rest
   B. intravenous hydralazine
   C. hydrochlorothiazide
   D. low-salt diet
   E. tranquilizers

Questions 52–55: A 28-year-old man presents to the emergency department with a two-hour history of chest tightness, coughing, and wheezing. The history determines that he has had many such attacks in recent years, usually brought on by emotional factors or exertion, and generally treatable by self-medication at home. There is a long history of hay fever and other members of the family have had similar symptoms.

Physical examination reveals dyspnea, orthopnea, and cyanosis. High pitched, sibilant rhonchi occur on inspiration and expiration and some coarse crepitations are audible. Pulse is 130/min and regular. Coughing produces viscid sputum. An emergency arterial PO_2 is 65 mmHg.

52. The most likely diagnosis is
   A. right heart failure
   B. left heart failure
   C. intrinsic asthma
   D. extrinsic asthma
   E. pneumonoconiosis

53. Precipitating factors in this condition include all of the following **EXCEPT**
   A. exposure to antigen
   B. salt ingestion
   C. excitement
   D. infection
   E. irritants

54. Pulmonary function studies during an acute attack are most likely to show
   A. decreased FEV_1
   B. increased vital capacity
   C. decreased residual volume
   D. decreased airway resistance
   E. increased arterial PO_2

55. The patient is best managed by
   A. fluid restriction
   B. cardioversion
   C. β-adrenergic drugs
   D. β-adrenergic blocker
   E. α-adrenergic drugs

Questions 56–59: An 18-year-old schoolboy develops tiredness and weakness, associated with a marked sore throat. There is no history of contact with others having the same symptoms. On review of systems he admits to chills and excessive sweating, but denies coryza, cough, nausea, vomiting, or diarrhea. On physical examination there is pharyngitis, marked cervical lymphadenopathy, fever of 38 °C, and the spleen is palpable 3 cm below the left costal margin. There is no hepatomegaly and no jaundice.

Initial laboratory data reveals a hemoglobin of 13 g%, white blood count of 13,000, and platelet count of 240,000. The differential white count includes 35% neutrophils, 5% bands, 50% lymphocytes, and 10% monocytes. Many of the lymphocytes have irregular or dark blue cytoplasm.

56. The diagnosis is most likely to be
   A. infectious mononucleosis
   B. infectious hepatitis
   C. streptococcal pharyngitis
   D. acute lymphocytic leukemia
   E. monocytic leukemia

57. All of the following antibodies are commonly seen **EXCEPT**
   A. anti-i cold agglutinins
   B. heterophil
   C. antismooth muscle
   D. anti-EB virus
   E. false positive VDRL

58. The causative agent in this disease is most likely
   A. *Streptococcus viridans*
   B. *Staphylococcus aureus*
   C. cytomegalovirus
   D. *Toxoplasma gondii*
   E. Epstein-Barr virus

59. The best short-term management of this patient includes
   A. digitalis
   B. high-dose steroids
   C. penicillin
   D. chlorambucil
   E. restriction of sports activities

Questions 60–63: A 34-year-old woman presents with a two-year history of the gradual onset of nervousness and fatigue. She has lost 15 lb in the past year despite an increase in appetite. Her menses have become scant, with frequent intermenstrual spotting, and she finds that she constantly prefers a cooler environment than those around her. On physical examination, she has a heart rate of 100/min with normal blood pressure, and a fine tremor of the fingers is noted. The skin is warm, moist, and has a smooth texture. The thyroid is diffusely enlarged to palpation. A lid-lag sign is easily demonstrable.

60. The laboratory testing is most likely to show
    A. high blood sugar
    B. decrease in serum T_3 and T_4
    C. increase in I^{131} uptake in thyroid
    D. a cold thyroid nodule
    E. high calcium, low phosphorus

61. Other common symptoms or signs in this condition include all of the following **EXCEPT**
    A. dyspnea
    B. palpitations
    C. thyroid bruit
    D. hepatomegaly
    E. swelling of the legs

62. A common complication of this disease is
    A. hepatitis
    B. interstitial pneumonia
    C. peptic ulcer disease
    D. infiltrative ophthalmopathy
    E. hemolytic anemia

63. The best management of this patient's disease might include all of the following **EXCEPT**
    A. propylthiouracil
    B. isoproterenol
    C. methimazole
    D. carbimazole
    E. radioiodine

Questions 64–67: A 28-year-old-woman presents with a history of diarrhea and crampy abdominal pain. In retrospect, the attacks have been increasing in frequency and severity for the past three years, so that there are now at least ten bowel movements per day. Attacks were frequently associated with fever. She has lost 10% of her body weight in the past year. On examination, she is a thin woman in acute distress from abdominal cramps. A boggy mass can be palpated in the right lower quadrant of the abdomen, associated with marked tenderness, but not guarding. Examination of the anal area reveals an apparent perianal fistula.

X-rays of the GI system are taken and on review show that a barium enema is essentially normal. An upper GI series shows a normal esophagus and stomach, but there are several areas of stenosis in the ileum separated by normal bowel. A mass of adherent bowel is seen in the right lower quadrant made up of adherent loops of ileum with evidence of fistulous connections between several loops.

64. The most likely diagnosis is
 A. ulcerative colitis
 B. Crohn's disease
 C. acute appendicitis
 D. celiac disease
 E. carcinoma of the small bowel

65. The pathology of this condition is most likely to reveal
 A. noncaseating granulomata
 B. caseating granulomata
 C. thinning of the affected bowel wall
 D. multiple free perforations of the ileum
 E. rectal inflammation

66. This long-standing disease may be associated with all of the following **EXCEPT**
 A. amyloidosis
 B. pyelonephritis
 C. polyarthritis
 D. emotional instability
 E. respiratory failure

67. Management of this condition may involve all of the following **EXCEPT**
   A. metronidazole
   B. sulfasalazine
   C. hydration
   D. p-aminosalicylic acid
   E. surgery

Questions 68–71: A 64-year-old man is admitted for investigation of symptoms of urinary frequency, hesitancy, and nocturia for the past six months. In addition, burning dysuria has occurred on two occasions, requiring treatment with antibiotics. He has a one-year history of angina pectoris, for which he takes occasional nitroglycerin. On physical examination, the blood pressure is 130/90, heart rate is 90/min and regular, and an enlarged prostate is palpable per rectum.

Laboratory data, including EKG, yield no contraindication to surgery and he is operated upon for a transurethral resection of the prostate. Anesthesia and surgery are uneventful and blood loss is minimal. Six hours postoperatively he experiences a shaking chill, a temperature of 40 °C, and his blood pressure is 90/60.

68. The most likely diagnosis is
   A. gram-negative bacteremia
   B. myocardial infarction
   C. postoperative bleeding
   D. arrhythmia
   E. lobar pneumonia

69. Further examination at the time of the hypotensive event is likely to show all of the following **EXCEPT**
   A. lactic acidosis
   B. cold, clammy skin
   C. elevated fibrin split products
   D. sinus tachycardia on EKG
   E. increased urine output

70. Urinary tract infection with subsequent bacteremia is most likely to be caused by
   A. *Salmonella typhimurium*
   B. *Shigella sonnei*
   C. *Vibrio cholerae*
   D. *Haemophilus influenzae*
   E. *E. coli*

71. The best management of this patient includes
   A. DC counter shock
   B. monitoring central venous pressure
   C. cardiac pacemaker
   D. salt restriction
   E. potassium infusion

DIRECTIONS (Questions 72–76): Each set of lettered headings below is followed by a list of numbered words or phrases. For each numbered word or phrase select
   A   if the item is associated with **A** only,
   B   if the item is associated with **B** only,
   C   if the item is associated with both **A** and **B**,
   D   if the item is associated with neither **A** nor **B**.

Questions 72–76:

   A. Low-output heart failure
   B. High-output heart failure
   C. Both
   D. Neither

72. Systemic vasoconstriction

73. Normal arterial venous oxygen difference

74. Cardiomyopathy

75. Arteriovenous fistula

76. Narrow pulse pressure

DIRECTIONS (Questions 77–89): Each group of questions below consists of lettered headings followed by a list of numbered words or statements. For each numbered word or statement, select the **one** lettered heading that is most closely associated with it. Each lettered heading may be selected once, more than once, or not at all.

Questions 77–81:

A. Nephrocalcinosis
B. Glycosuria
C. Tufting and mushrooming of terminal phalanges
D. Idiopathic destructive atrophy
E. Elevated cholesterol

77. Hypothyroidism

78. Cushing's syndrome

79. Addison's disease

80. Acromegaly

81. Hyperparathyroidism

Questions 82–85:

A. Farmer's lung
B. Humidifier lung
C. Silicosis
D. Bagassosis
E. Furrier's lung

82. *Penicillium* species

83. *Thermoactinomyces sacchari*

84. *Aspergillis fumigatus*

85. Kaolin

Questions 86–89:

    A. Poststreptococcal glomerulonephritis
    B. Idiopathic nephrotic syndrome
    C. Systemic lupus erythematosis
    D. Goodpasture's syndrome

86. Diffuse epithelial foot process effacement

87. Antibodies to type IV collagen

88. Prolonged decrease in C3 component of complement

89. Transient decrease in C3 component of complement

DIRECTIONS (Questions 90–118): For each of the questions or incomplete statements below, **one** or **more** of the answers or completions given is correct. Select
    A   if only **1, 2,** and **3** are correct,
    B   if only **1** and **3** are correct,
    C   if only **2** and **4** are correct,
    D   if only **4** is correct,
    E   if **all** are correct.

---

90. Which of the following drugs used in the treatment of hypertension act as β-adrenergic blockers?
    1. Chlorthalidone
    2. Clonidine
    3. Minoxidil
    4. Metoprolol

91. Which of the following techniques are useful in estimating the size of a myocardial infarction?
    1. Creatine kinase enzyme measurement
    2. Precordial electrocardiographic mapping
    3. "Hot-spot" scintigraphy
    4. "Cold-spot" scintigraphy

Directions Summarized				
A	**B**	**C**	**D**	**E**
1,2,3	*1,3*	*2,4*	*4*	*All* are
only	only	only	only	correct

92. Digital subtraction angiography offers the following advantages over standard angiography.
    1. Enhancement of contrast opacification of the blood pool
    2. Facilitation of quantitative analysis
    3. Use of lower doses of contrast media
    4. Reduced allergic reaction to contrast media

93. Carotid sinus massage leads to abrupt slowing of tachycardia in the following arrhythmias.
    1. Sinus tachycardia
    2. Type II second degree AV block
    3. Right bundle branch block
    4. Reciprocating tachycardia using an accessory pathway

94. Which of the following antihypertensive drugs inhibits conversion of angiotensin I to angiotensin II?
    1. Minoxidil
    2. Captopril
    3. Diltiazem
    4. Enalapril

95. Which of the following factors influences the incidence of multiple sclerosis?
    1. Geographic location
    2. Occurrence in a spouse
    3. Occurrence in a first-degree blood relative
    4. Exposure to aluminum

96. Pathological changes in Alzheimer's disease include
    1. widened cerebral sulci
    2. enlargement of the third ventricle
    3. silver-staining neurofibrillary masses
    4. degeneration of pyramidal cells in the hippocampus

97. Amyotrophic lateral sclerosis is best characterized by
1. early loss of the fine finger movements
2. abnormal muscle cramps
3. fasciculations of the shoulder muscles
4. loss of tendon reflexes

98. Guillain-Barré syndrome is most likely to include
1. primarily females
2. early cranial nerve involvement
3. prominent muscle atrophy
4. disturbances of autonomic function

99. Hypocalcemia in chronic renal failure may result from
1. calcium entry into bone
2. impaired synthesis of vitamin D metabolites
3. increased phosphate retention
4. failure of PTH function

100. Which of the following lesions is characteristic of diabetic nephropathy?
1. Glomerulosclerosis
2. Arterionephrosclerosis
3. Chronic interstitial nephritis
4. Papillary necrosis

101. Chronic bacterial prostatitis may be best characterized as
1. associated with bladder infection
2. responsive to most antibiotics
3. resistant to transurethral prostatectomy
4. a form of venereal disease

102. Fibromuscular dysplasias of the renal artery are
1. frequently bilateral
2. tenfold more common in females
3. more easily resected than atherosclerotic plaques
4. associated with diabetes mellitus

		Directions Summarized		
A	**B**	**C**	**D**	**E**
1,2,3	*1,3*	*2,4*	*4*	*All* are
only	only	only	only	correct

103. Idiopathic hypercalciuria leading to calcium stones may be characterized as
  1. occurring sporadically in the population
  2. absorbing increased calcium in the proximal tubules
  3. having decreased activation of 1,25-dihydroxyvitamin D
  4. responding to thiazide diuretics

104. Spherocytes seen on a peripheral blood film may indicate
  1. hemoglobin C disease
  2. antibodies on red cells
  3. erythroleukemia
  4. an autosomal dominant condition

105. Refractory anemia with excess blasts
  1. is generally benign
  2. achieves remission after treatment with granulocyte growth factors
  3. improves with iron therapy
  4. may show Auer rods

106. The best policy for breast cancer screening includes
  1. annual breast self-examination
  2. baseline mammogram between ages 35 and 40
  3. cytologic examination of breast milk postpartum
  4. mammography and physical exam annually for all women over 40

107. Risk factors for carcinoma of the esophagus include
  1. male sex
  2. country of origin
  3. alcohol
  4. tobacco

108. Biologic therapy with lymphokine-activated cells plus inter-leukin-2, or with tumor-infiltrating lymphocytes has shown promise in treating
   1. renal cell carcinoma
   2. acute myeloblastic leukemia
   3. malignant melanoma
   4. colorectal carcinoma

109. Disseminated testicular cancer is treated at presentation with various combination chemotherapy regimens, all of which contain
   1. adriamycin
   2. bleomycin
   3. actinomycin D
   4. cisplatinum

110. Serologic studies of a patient with recent onset of acute hepatitis B are likely to show
   1. positive hepatitis B surface antigen
   2. IgG antibodies to HBc
   3. circulating hepatitis Be antigen
   4. antibodies to hepatitis Bs

111. The irritable bowel syndrome may be characterized by
   1. lactase deficiency
   2. increased resting colonic motility
   3. melena
   4. decreased resting colonic motility

112. Treatment of chronic gouty arthritis in patients who are allergic to allopurinol may make use of
   1. probenicid
   2. high fluid intake
   3. acetazolamide
   4. sulfinpyrazone

Directions Summarized				
A	**B**	**C**	**D**	**E**
1,2,3	*1,3*	*2,4*	*4*	*All* are
only	only	only	only	correct

113. Endocrine abnormalities of anorexia nervosa often include
    1. increased follicle-stimulating hormone
    2. decreased total T4 levels
    3. decreased plasma cortisol levels
    4. decreased leutinizing hormone

114. Which of the following therapies are appropriate for management of thyroid storm?
    1. Propylthiouracil
    2. H-2 blockers
    3. Corticosteroids
    4. Hydralazine

115. Gynecomastia may be associated with
    1. hepatic cirrhosis
    2. renal failure
    3. thyrotoxicosis
    4. cimetidine therapy

116. Patients with early manifestations of Lyme disease are likely to have
    1. a red macule with central clearing
    2. multiple annular secondary skin lesions
    3. fever
    4. diarrhea

117. Reiter's syndrome is typically associated with
    1. a self-limiting course
    2. nonspecific urethritis
    3. females
    4. buccal ulceration

118. Which of the following agents frequently can be linked to erythema multiforme syndrome?
   1. Mycoplasma pneumoniae
   2. Penicillin
   3. Hydantoins
   4. Sulfonamides

Explanatory Answers

1. B. The echocardiographic hallmark of mitral stenosis is the absence of valve closure in mid-diastole and of reopening in late-diastole. A decreased diastole slope is characteristic but not specific. (Ref. 5, p. 103)

2. A. The pulmonary artery or capillary wedge pressure has a wave form similar to that of left atrial pressure, but is clamped and delayed by transmission through the capillary vessels. (Ref. 5, p. 250)

3. E. Pulmonic stenosis is frequently asymptomatic. The intensity of the click decreases with inspiration. (Ref. 5, p. 991)

4. C. The patient has prosthetic endocarditis, which is associated with anemia in 50% to 80% of patients and may be the first finding to bring the diagnosis to attention. (Ref. 5, p. 1112)

5. A. The patient has hyperkalemia with a potassium level of 8.2 mg%. On the EKG, no atrial activity is detected. The ventricular rate is slightly irregular. Beat number 4 is a ventricular premature contraction. The T waves are tall and markedly peaked. (Ref. 1, p. 547)

6. C. Interventional management of angina at age 65 is reserved for angina at rest or unstable angina refractory to front-line medical management. (Ref. 5, p. 1342)

7. B. The histology of the pericardium in Dressler syndrome reveals a nonspecific inflammation that is diffuse in contrast to the patchy distribution in viral pericarditis. (Ref. 5, p. 1521)

8. C. Ventricular aneurysms usually develop within days of myocardial infarction and enlarge over subsequent weeks or months. They may be associated with arrhythmias, congestive heart failure, mural thrombosis, and systemic embolization. (Ref. 1, p. 333)

9. B.   The patient has contracted HIV infection from shared needles and has common presenting symptoms of AIDS. (Ref. 4, p. 1671)

10. D.   This subject remains controversial, but most authors accept the beneficial effects of ASA in males. Blood pressure should not be lowered, and only extracranial surgery is advocated by some. (Ref. 3, p. 603)

11. A.   The histiocytosis X syndrome includes eosinophilic granuloma, and frequently involves the pituitary leading to diabetes insipidus. Other features include lytic bone lesions and the reticulonodular lung infiltrate seen on the x-ray. (Ref. 1, p. 430)

12. D.   Next to nystagmus, the most frequent ocular abnormality is a lateral rectus weakness which is always bilateral but not necessarily symmetrical. (Ref. 3, p. 762)

13. E.   Many side effects of L-dopa are troublesome and include nausea, depression, and the induction of involuntary movements. (Ref. 3, p. 878)

14. C.   Lesions may increase during pregnancy. About one-third of patients present with neurologic symptoms, and eighth nerve tumors are common. (Ref. 3, p. 918)

15. D.   The patient is having a cholinergic crisis. Other symptoms include pallor, salivation, colic, diarrhea, and miosis. (Ref. 3, p. 1082)

16. B.   The Z gene clearly leads to COPD, but other abnormal genes have been described. Hepatic manifestation may also arise. (Ref. 1, p. 415)

17. A.   Recombinant techniques for producing pure erythropoietin have led to trials demonstrating responsiveness to this growth factor. (Ref. 1, p. 571)

18. E.   The number of units that are optimal is not clear, but most studies suggest that five units would be optimal. (Ref. 2, p. 1166)

19. B.   The patient has sickle cell anemia, which is a hereditary defect in the β chain of hemoglobin. The biconcave appearance of the vertebral bodies is characteristic. Some sclerotic changes are also seen. (Ref. 6, p. 376)

20. C.   Heroin abuse may lead to focal and segmental glomerulosclerosis, leading to nephrotic syndrome and renal failure; but amphetamine abuse causes vasculitis. (Ref. 2, p. 1189)

21. E.   The radiographic picture shows a characteristic worm-eaten appearance with tortuous thickened esophageal folds. The left anterior oblique projection is most ideal for its demonstration. (Ref. 2, p. 1346)

22. D.   Most strains are sensitive to many antibiotics. Amoxicillin and trimethoprim-sulfamethoxazole have been used most widely. (Ref. 2, p. 1192)

23. A.   This common condition is found in 1 in 500 autopsies. Inheritance is autosomal dominant. Hepatic cysts are present in 30% of patients. (Ref. 2, p. 1205)

24. B.   Patients under 40 with acute myeloblastic leukemia are optimal candidates for bone marrow transplant from a matched sibling donor. (Ref. 6, p. 658)

25. B.   Acute leukemia is the most common second malignancy, followed by non-Hodgkin's lymphoma. (Ref. 6, p. 881)

26. E.   DDAVP releases stored factor VIII in patients with mild Von Willebrand disease and is usually safe and sufficient for controlling bleeding. (Ref. 6, p. 1118)

27. C.   Most syndromes are associated with small cell carcinoma. Ectopic PTH leads to hypercalcemia through physiological activity of the aberrant peptide. (Ref. 7, p. 611)

28. D.   Thymomas have a number of syndromes associated with them, including myasthenia gravis, and aplasias of the marrow. (Ref. 7, p. 714)

29. A. Local failure and spread to the liver are the most common therapeutic problems. Distant metastases are relatively uncommon (less than 30%). (Ref. 7, p. 770)

30. B. α-Fetoprotein is the major tumor marker and is elevated in over 70% of patients with disease. High levels carry a poor prognosis. (Ref. 7, p. 840)

31. B. This enzyme is responsible for maintaining the intracellular folates of plates in a reduced state, required for thymidylate synthesis. (Ref. 7, p. 350)

32. A. Adjuvant chemotherapy imparts no significant survival advantage to postmenopausal women, whereas the effects of tamoxifen are well documented. (Ref. 7, p. 1233)

33. D. The clinical presentation of sarcoid is often asymptomatic with abnormal chest x-ray. Anorexia, weight loss, and fever are usually absent. (Ref. 1, p. 451)

34. E. Vitamin K absorption from the GI tract is reduced, leading to decreased synthesis of factors II, VII, IX, and X. (Ref. 2, p. 1313)

35. C. Biopsy of the bowel mucosa demonstrates PAS-positive macrophages in the lamina propria and electron microscopic evidence of bacilliform bodies. (Ref. 2, p. 1270)

36. D. Adherence of sucralfate to granulation tissue may prevent diffusion of hydrochloric acid to the base of the ulcer. (Ref. 2, p. 1245)

37. C. Mild hereditary glucuronyl transferase deficiency, or Gilbert's syndrome, typically shows fluctuating jaundice exacerbated by fasting, surgery, or infection. (Ref. 2, p. 1321)

38. B. Gaucher's disease is a relatively common familial disorder characterized by accumulation of glucocerebroside within phagocytic cells. (Ref. 1, p. 1145)

39. E. Middle-aged men with very high LDL cholesterol should receive a bile acid-binding resin, high dose nicotinic acid, in combination with clofibrate, or gemfibrozil. (Ref. 1, p. 1144)

40. A. The patient has porphyria cutanea tarda, a relatively common disorder leading to iron damage to the liver, best managed by intermittent phlebotomy. (Ref. 1, p. 1186)

41. D. The hallmark of diabetes insipidus is urine specific gravity less than 1.005 and urine osmolality less than 200. (Ref. 1, p. 1310)

42. A. Non-insulin-dependent diabetes usually occurs over age 40, rarely has ketosis, and does not have circulating islet cell antibodies. (Ref. 1, p. 1366)

43. C. If not well controlled, these patients will have low serum calcium, high serum phosphorus, and high PTH. (Ref. 1, p. 1483)

44. D. Serum sodium and potassium levels are usually normal, while bicarbonate levels are often depressed, possibly due to lactic acidosis. (Ref. 1, p. 1377)

45. A. The American Rheumatism Association criteria for definite rheumatoid arthritis include all the points in this patient's clinical presentation. (Ref. 1, p. 1999)

46. D. Although all of the tests listed may be abnormal, antibodies to native DNA are the most reliable, with antinuclear tests next most useful. (Ref. 1, p. 2016)

47. D. This therapy, called PUVA, utilizes oral psoralens followed by long-wave ultraviolet light, which activates the psoralens in the affected areas. (Ref. 2, p. 233)

48. E. The presence of bacteriuria would be potentially indicative of pyelonephritis; but proteinuria may occur with either renal hypertension or essential hypertension with accompanying renal function impairment. (Ref. 1, p. 277)

49. B. The intravenous pyelogram frequently provides an important clue to clinically unsuspected renal disorders. Pelvocalyceal structure, length, and appearance time of contrast media are all important. (Ref. 1, p. 629)

50. A. It is noteworthy that a single normal value for serum potassium does not necessarily exclude any of these hypertensive diseases. (Ref. 1, p. 277)

51. B. Available evidence indicates that neurogenic, endocrine, renal, vascular, hormonal, and probably other mechanisms contribute components of variable magnitude in individual patients. (Ref. 1, p. 277)

52. D. Features of extrinsic asthma include childhood onset, family history of allergy, hay fever, positive skin tests, and raised IgE levels. (Ref. 1, p. 403)

53. B. The attack may begin within minutes of exposure to antigen and may or may not be associated with symptoms of hay fever. (Ref. 1, p. 404)

54. A. Any elevation of the PCO_2 is of grave significance, for it indicates that the resting ventilation is maximal and that reduction through fatigue may be lethal. (Ref. 1, p. 407)

55. C. Acute exacerbations of asthma may be promptly relieved by inhalation of isoproterenol or by epinephrine subcutaneously. (Ref. 1, p. 408)

56. A. Most patients exhibit malaise, sweats, sore throat, and anorexia; but diarrhea is rare. Splenomegaly occurs in half the patients; hepatomegaly is uncommon. (Ref. 6, p. 538)

57. C. The heterophil antibody is an IgM antibody which reacts with sheep red cells. Activity is absorbed out by beef red cells, but not by guinea pig kidney. (Ref. 6, p. 542)

58. E. Typical heterophil-positive infectious mononucleosis is almost always associated with a rise in the titre of antibodies to EBV-related antigens. (Ref. 6, p. 542)

59. E. The spleen is often fragile and subject to rupture. Fatigue will be exacerbated by exertion. (Ref. 6, p. 539)

60. C. Hypersecretion of hormone leads to an increase in the concentrations of serum T_4 and T_3, but the T_3 may be disproportionately increased. (Ref. 1, p. 1321)

61. D. In older patients, circulatory manifestations may predominate, while nervous manifestations may be lacking. (Ref. 1, p. 1321)

62. D. This eye complication occurs in about 50% of patients with Grave's disease and is not uncommonly independent of the thyrotoxic aspect. (Ref. 1, p. 1324)

63. B. All major forms of treatment exert their effects by imposing restraints on the rate of hormone secretion, so that therapy is essentially palliative. (Ref. 1, p. 1321)

64. B. The nature of the onset of symptoms varies greatly and depends on the anatomic site and longitudinal areas of involvement. (Ref. 2, p. 1280)

65. A. It is generally agreed that the examination of surgical specimens will show granulomas in over two-thirds of patients. (Ref. 2, p. 1280)

66. E. The polyarthritis usually described in regional enteritis is often only of several weeks' duration, asymmetrical, migratory, and subacute. (Ref. 2, p. 1286)

67. D. In a multicentre study, sulfasalazine was shown to be effective in therapy of active colonic Crohn's disease. Corticosteroids were also effective, but less so than with small bowel disease. (Ref. 2, p. 1288)

68. A. Urinary tract infection accounts for about two-thirds of all cases of blood invasion by the enteric bacteria. (Ref. 4, p. 1226)

69. E.   Lactic acidosis is common in severely ill patients. There is a marked anion gap with low levels of serum bicarbonate and PCO_2. (Ref. 4, p. 456)

70. E.   Enteric gram-negative bacteria probably arising from the gastrointestinal tract are the common pathogens, and include *E. coli,* the *Klebsiella-Enterobacter* group, and *Proteus* species. (Ref. 4, p. 460)

71. B.   Management also includes measures designed to improve cardiac function, tissue perfusion, and electrolyte imbalance, especially acidosis. Use of high-dose steroids is controversial. (Ref. 4, p. 468)

72. A.   Low-output heart failure is characterized by clinical evidence of impairment of the peripheral circulation with vasoconstriction. In high-output failure, the extremities are warm. (Ref. 5, p. 473)

73. B.   The arterial-venous oxygen difference is abnormally widened in low-output failure, but is normal or low in high-output failure. (Ref. 5, p. 473)

74. A.   Low cardiac output at rest characterizes failure occurring in most forms of heart disease, including congenital, valvular, rheumatic, and cardiomyopathic. (Ref. 5, p. 473)

75. B.   High-output states that lead to failure include thyrotoxicosis, arteriovenous fistulas, beriberi, Paget's disease of bone, anemia, and pregnancy. (Ref. 5, p. 473)

76. A.   In low-output failure, the stroke volume declines, leading to narrow pulse pressure. In high-output failure, the pulse pressure is normal or widened. (Ref. 5, p. 473)

77. E.   Both the synthesis and the degradation of lipids are depressed, with the net result being lipid synthesis. (Ref. 1, p. 1328)

78. B.   The usual pattern is not that of frank diabetes mellitus, but rather, a failing of blood glucose to return to fasting levels appropriately. (Ref. 1, p. 1353)

79. D. The former leading cause was destruction of glands by tuberculosis, but is now idiopathic atrophy. (Ref. 1, p. 1300)

80. C. Articular cartilage proliferation initially widens the joint space, but as the disease progresses, cartilage may become eroded. (Ref. 1, p. 1351)

81. A. Manifestations include copious excretion of dilute urine and, in later stages, uremia, edema, and hypertension. (Ref. 1, p. 1488)

82. B. Humidifier lung may be associated with air conditioning systems, and is caused by organisms that are thermotolerant, including *T. vulgaris, T. candidus,* and various bacteria and protozoa. (Ref. 1, p. 424)

83. D. Bagassosis follows inhaled organic dust exposure and leads to interstitial lung disease from inhaled *T. sacchari.* (Ref. 1, p. 424)

84. A. Farmer's lung results from exposure to moldy hay. Patients present 4 to 8 hours after exposure with fever, chills, and cough. (Ref. 2, p. 1072)

85. C. Silicosis results from occupational exposure in mining, quarrying, and blasting. Acute and chronic syndromes are recognized. (Ref. 2, p. 1070)

86. B. There is little or no alteration of the glomerular capillaries in minimal lesion nephrotic syndrome. (Ref. 2, p. 1178)

87. D. Circulating antibodies to glycopeptide antigens are found in over 90% of cases if serums are examined early in the course of disease. (Ref. 2, p. 1185)

88. C. Depressed levels of C_3 and C_4 are seen in the untreated patient, as well as high levels of precipitating and nonprecipitating antibody to double-stranded DNA. (Ref. 2, p. 1184)

89. A. C_3 complement is transiently decreased and returns to normal within eight weeks after the first signs of renal disease. (Ref. 2, p. 1174)

90. D. β-Adrenergic blocking drugs used in the treatment of hypertension include propranolol, oxprenolol, atenolol, and metoprolol. The mechanisms of hypotensive action include reduction of cardiac output and loss of the full compensatory increase in peripheral resistance. (Ref. 1, pp. 276–280)

91. E. All techniques listed are useful in estimating infarct size. Since none is perfect in isolation, two or more are usually used at the same time. All are noninvasive. (Ref. 1, p. 295)

92. A. Digital angiography has evolved as a technique for enhancing contrast of angiograms performed with direct ventricular or coronary injection of low doses of contrast media and for sophisticated analysis of data. (Ref. 5, p. 356)

93. D. Carotid sinus massage leads to gradual slowing in sinus tachycardia, type II AV block, and right bundle branch block. Abrupt slowing occurs with atrial flutter and may occur in reciprocating tachycardia. (Ref. 5, p. 660)

94. C. Minoxidil is a vasodilator and diltiazem is a calcium antagonist. Captopril and enalopril inhibit conversion of the inactive decapeptide of angiotension to the active octapeptide. (Ref. 5, p. 877)

95. B. Several authors have shown that the incidence differs markedly at different latitudes, lowest at the equator, and highest in Northern Europe and North America. (Ref. 3, p. 701)

96. E. All are characteristic. In addition, there are plaques composed of amorphous materials scattered through the cortex. (Ref. 3, p. 864)

97. A. The tendon reflexes are notable for their liveliness. The disease starts in the upper extremities and spreads to the neck, tongue, trunk, and lower extremities. (Ref. 3, p. 890)

98. D. Disturbance of autonomic function are common, and include sinus tachycardia, facial flushing, and fluctuating blood pressure. (Ref. 3, p. 968)

99. A. Increased phosphate retention leads to calcium entry into bone and elevated levels of PTH. Vitamin D_3 is reduced. (Ref. 2, p. 1159)

100. E. Diabetic nephropathy is associated with a variety of clinical syndromes, including mild proteinuria, nephrotic syndrome, and acute or chronic renal failure. (Ref. 2, p. 1187)

101. B. Chronic recurrent bladder infections are associated. The condition responds poorly to antibiotics; only one-third are cured by transurethral resection. (Ref. 2, p. 1195)

102. A. The lesions are more easily resected with lower morbidity presumably because the individuals are younger than those with atherosclerosis. (Ref. 2, p. 1201)

103. D. The condition is inherited. The renal tubules absorb less calcium and renal activation of vitamin D is increased. Thiazide diuretics lower urine calcium and prevent formation of stones. (Ref. 2, p. 1213)

104. B. A positive Coomb's test indicating antibodies on red cells is often associated with spherocytes, as is hereditary spherocytosis. (Ref. 6, pp. 237, 281)

105. D. This form of myelodysplastic syndrome usually progresses to acute myeloblastic leukemia, and may have cytoplasmic rods. (Ref. 6, p. 613)

106. C. Breast self-examination should be frequent and should be reinforced by a physician or nurse. Mammography technique is critical. (Ref. 7, p. 483)

107. E. The roles of alcohol and tobacco are clearest in North America and Europe, but other environmental factors are suspect in Asia. (Ref. 7, p. 727)

108. B. Most patients with metastatic renal cell carcinoma have had the primary tumor resected before receiving this adoptive immunotherapy. (Ref. 7, p. 993)

109. C. The addition of cisplatinum to the chemotherapy treatment has dramatically improved the long-term survival rate to the 70% range. (Ref. 7, p. 1086)

110. B. Positive test for circulating HBs antigen combined with IgM anti-Hbc antibodies and circulating HBe antigen indicates acute HBV infection with high infectivity. (Ref. 2, p. 1331)

111. C. The irritable bowel syndrome may present with either diarrhea or constipation, the so-called spastic colon variant. (Ref. 2, p. 1294)

112. E. All may be used. Probenecid and sulfinyprazone begin to lose effectiveness when the creatinine clearance falls, and are inhibited by salicylates. (Ref. 1, p. 1170)

113. D. Male anorectics lose libido. Patients have decreased FSH and LH, decreased T3, normal T4, and normal or slightly elevated cortisol. (Ref. 1, p. 1216)

114. B. Other therapeutic interventions include intravenous fluids, sodium iodide, acetaminophen, propranalol, and oxygen. (Ref. 1, p. 1327)

115. E. Hepatic cirrhosis causes increased estrogen production, whereas renal failure leads to androgen deficiency, resulting from testicular failure. (Ref. 1, p. 1411)

116. A. Early symptoms of Lyme disease include malaise, headache, fever, chills, stiff neck, arthralgias, and rash; but only 2% have diarrhea. (Ref. 1, p. 1727)

117. C. Reiter's syndrome most commonly presents in males with urethritis, conjunctivitis, and arthritis; buccal ulcerations are usually seen with careful inspection. (Ref. 1, p. 2007)

118. E. Other agents linked to erythema multiforme are antipyretics and barbiturates. About 50% of cases have no etiology determined. (Ref. 2, p. 238)

References

1. Wyngaarden JB, Smith LH Jr: *Cecil Textbook of Medicine,* ed 18. Philadelphia, WB Saunders Co, 1988.

2. Braunwald E, Isselbacher KJ, Petersdorf RG, Wilson JD, Martin JB, Fauci, AS: *Harrison's Principles of Internal Medicine,* ed 11. New York, McGraw-Hill Book Co, 1987.

3. Adams RD, Victor M: *Principles of Neurology,* ed 3. New York, McGraw-Hill Book Co., 1985.

4. Mandell GL, Douglas RG Jr, Bennett JE: *Principles and Practice of Infectious Diseases,* ed 2. New York, John Wiley and Sons, 1985.

5. Braunwald E: *Heart Disease,* ed 3. Philadelphia, WB Saunders Co, 1988.

6. Jandl JH: *Blood.* Boston, Little, Brown and Co, 1987.

7. DeVita VT, Hellman S, Rosenberg S: *Cancer,* ed 3. Philadelphia, JB Lippincott Co, 1989.

2 Pediatrics

Victor LaCerva, MD

DIRECTIONS (Questions 119–138): Each of the questions or incomplete statements below is followed by five suggested answers or completions. Select the **one** that is best in each case.

---

119. Which of the following is **NOT** a characteristic of early congenital syphilis?
   A. Snuffles
   B. Saber shin
   C. Pseudoparalysis
   D. Syphilitic nephrosis
   E. Cranial bossing

120. True statements about tuberculosis in children include each of the following **EXCEPT**
   A. the source is usually an adult member of the household
   B. a positive PPD is induration of greater than 5 mm
   C. BCG vaccination is not recommended in this country
   D. there are no characteristic hematologic changes
   E. hospitalization is usually not indicated

121. A child previously immunized with inactivated measles vaccine
   A.   will not develop wild measles infection
   B.   may develop severe atypical measles on exposure to wild virus
   C.   develops secretory IgA against measles
   D.   may develop encephalitis on revaccination
   E.   should receive yearly boosters

122. Most neonatal herpes is
   A.   localized to the skin
   B.   not associated with systemic symptoms
   C.   caused by type I virus
   D.   caused by type II virus
   E.   usually benign

123. Acute herpetic gingivostomatitis is **NOT** associated with
   A.   fever
   B.   swollen gums
   C.   lymphadenopathy
   D.   ulcers of the buccal mucosa
   E.   papular rash

124. The most common complication of chickenpox in children is
   A.   pneumonia
   B.   encephalitis
   C.   cystitis
   D.   angioneurotic edema
   E.   secondary bacterial skin infection

125. Which of the following is **NOT** a characteristic of rubella?
   A.   Rash that typically begins on the trunk
   B.   Retroauricular and postoccipital adenopathy
   C.   Low-grade fever
   D.   Arthralgia
   E.   Splenomegaly

126. Which of the following statements concerning roseola infantum is true?
A. The etiology is believed to be due to Coxsackie virus
B. Fevers are generally quite high at the onset of the illness
C. The rash often begins with a slapped cheek appearance
D. The illness is not considered to be contagious
E. The rash leaves temporary hypopigmented areas after fading

127. Pyloric stenosis is characterized by all of the following **EXCEPT**
A. it is more common in males
B. it presents with persistent vomiting at 3 to 6 weeks of age
C. may be accompanied by hyperbilirubinemia
D. it should always be confirmed by barium swallow
E. chalasi of the esophagus and hiatal hernia are differential diagnoses

128. All of the following statements concerning circumcision are true **EXCEPT**
A. it should not be routinely recommended for all newborns
B. it can have significant complications such as bleeding and infection
C. it is contraindicated in hypospadias and ambiguous genitalia
D. the newborn foreskin normally adheres to the glans
E. it is protective against the development of penile cancer

129. Which statement is true concerning Hodgkin's disease?
A. The peak incidence is between 15 and 34 years of age
B. The disease is twice as common in boys as girls
C. More than 90% of patients initially achieve a complete clinical remission
D. The most common presenting finding is enlargement of cervical lymph nodes
E. All of the above

130. All of the following statements about orthostatic proteinuria are true **EXCEPT**
    A. most children are healthy and there is no underlying pathology
    B. there is proteinuria in the recumbent position which decreases in the upright position
    C. the sex incidence is equal
    D. the usual age of detection is during the second decade
    E. there is no risk for the development of hypertension

131. Causes of hematuria in children include all of the following **EXCEPT**
    A. Wilms' tumor
    B. minimal lesion nephrotic syndrome
    C. polycystic kidneys
    D. subacute bacterial endocarditis
    E. renal artery thrombosis

132. Insulin requirements are increased in all of the following conditions **EXCEPT**
    A. acute infections
    B. onset of puberty
    C. emotional stress
    D. development of antibodies against insulin
    E. associated hypothyroidism

133. Acquired hypothyroidism most often results from
    A. cystinosis
    B. carcinoma of the thyroid
    C. lymphocytic thyroiditis
    D. hypopituitarism
    E. excision of a thyroglossal duct cyst

134. Sickle cell disease
    A. is found only in blacks
    B. usually manifests in the first few months of life
    C. results in an increased susceptibility to staph infections
    D. requires no therapy except during acute episodes
    E. results in a hemoglobin concentration of 10 to 12 g/dL

135. The following are all common causes of intoeing **EXCEPT**
   A. metatarsus adductus
   B. internal tibial torsion
   C. genu valgum
   D. anteversion of the hip
   E. transitory positional foot

136. Leukocoria is seen in all of the following conditions **EXCEPT**
   A. retrolental fibroplasia
   B. retinoblastoma
   C. cataract
   D. toxocara cyst
   E. iridocyclitis

Table 2.1 Glasgow Coma Scale

Subscale	Response	Score
Best eye opening	Spontaneous	4
	To voice	3
	To pain	2
	None	1
Best verbal response	Oriented	5
	Confused conversation	4
	Inappropriate words	3
	Incomprehensible sounds	2
	None	1
Best motor response, upper limb	Obeys commands	6
	Localizes pain	5
	Flexor withdrawal (decorticate posturing)	4
	Abnormal flexion (decerebrate posturing)	3
	Extension	2
	Flaccid	1

137. When a child presents in coma, it is important to document the degree. What score on the Glasgow Coma Scale (Table 2.1) suggests significant compromise of brain function?
    A. 5 or below
    B. 6 or below
    C. 7 or below
    D. 8 or below
    E. 9 or below

138. In the under-five-years-old age group, the most common malignancy is
    A. brain tumor
    B. lymphoma
    C. neuroblastoma
    D. acute leukemia
    E. eye malignancy

DIRECTIONS (Questions 139–163): This section consists of clinical situations, each followed by a series of questions. Study each situation and select the **one** best answer to each question following it.

Questions 139–143: A six-year-old boy, with no significant past medical history, appears at your office. He has recently recovered from an upper respiratory infection, and since early this morning has had a number of episodes of vomiting. His mother states that he seems somewhat sleepy this morning as well. He has minimal fever, and no obvious site of infection. He vomits after your exam, and you are concerned with his degree of lethargy.

139. Important points to note in the history include all of the following **EXCEPT**
    A. history of having taken aspirin recently
    B. level of influenza activity in the community
    C. history of head trauma
    D. history of renal infection
    E. length of time since the vomiting began

140. The most appropriate action at this time is
  A. immediate spinal tap
  B. admit for observation; obtain liver profile
  C. take throat, urine, and blood cultures; send home with instructions to keep you informed of the child's condition
  D. start on penicillin for presumed strep throat, after taking a throat culture
  E. obtain CT scan and neurosurgical consult

141. All of the following are true concerning Reye syndrome **EXCEPT**
  A. it can result in coma and death due to liver and brain mitochondrial dysfunction
  B. it is characterized by normal cerebrospinal fluid except for elevated pressure
  C. it is epidemiologically associated with influenza, chickenpox, and aspirin ingestion
  D. a liver biopsy is necessary for a correct diagnosis
  E. deepening of coma is most likely when the serum ammonia level is high

142. Which of the following is **NOT** characteristic of Reye syndrome?
  A. Delirium and stupor occur simultaneously or within a few hours of the onset of vomiting
  B. Proteinuria and jaundice
  C. Slight liver enlargement
  D. Increased concentration of plasma free fatty acids
  E. Hypoprothrombinemia

Table 2.2 Clinical Staging of Reye Syndrome

Grade	Symptoms at Admission
I	Quiet, lethargic, and sleepy, with vomiting and evidence of liver dysfunction
II	Deep lethargy, confusion, combative, hyperventilation, and hyperreflexic
III	Obtunded, light coma, decorticate rigidity, intact pupillary light reaction
IV	Seizures, deepening coma, decerebrate rigidity, fixed pupils
V	Coma, loss of tendon reflexes, respiratory arrest, fixed dilated pupils, isoelectric EEG

143. Successful management of Reye syndrome (Table 2.2) requires each of the following **EXCEPT**
   A. close observation when the patient is in stage I
   B. administration of glucose IV to avoid glycogen depletion
   C. restriction of fluids to 2500 mL/m2/d when cerebral edema is present
   D. close intracranial pressure monitoring in stages IV and V
   E. osmotherapy with mannitol in the more severely ill

Questions 144–148: You are called to the nursery to examine a newborn who is slightly jaundiced on the second day. Your examination is normal, and there is nothing significant in the history except that the mother received essentially no prenatal care. She did have a test for syphilis, which was negative; she was immune to rubella; and there is no blood group incompatibility. Upon further questioning of the mother, she admits to having had occasional episodes of "recreational" IV drug use during pregnancy.

144. This infant may be at risk for all of the following **EXCEPT**
   A. future CNS malignancy
   B. AIDS
   C. hepatitis B
   D. congenital malformations
   E. withdrawal manifestations

145. Which of the following is true about hepatitis B?
    A. The incubation period is 2 to 5 months
    B. Breastfeeding by mothers with positive HB(s)Ag is contraindicated
    C. There is an increased risk of abortion or malformations following hepatitis during the first trimester
    D. There is a vaccine available for hepatitis B, but it carries an extremely minimal risk of AIDS
    E. Ingestion of contaminated food or water is a common mode of transmission

146. Differential diagnosis of hepatitis B in the newborn period includes all of the following **EXCEPT**
    A. bacterial infection
    B. physiologic jaundice
    C. hemolytic disease
    D. homocystinuria
    E. biliary atresia

147. True statements about transmission of hepatitis B from pregnant carriers to their infants include all of the following **EXCEPT**
    A. hepatitis B tends to be milder in infants and children and tends to go unrecognized
    B. only about 50% of infants born to positive mothers will be positive for HB(s)Ag at birth
    C. the presence of HB(e)Ag in maternal carriers correlates highly with transmission of hepatitis B infection to their offspring
    D. although newborns are rarely symptomatic, most of those infected will go on to become carriers
    E. the risk of chronic liver disease or hepatocellular carcinoma is increased in these infants

Condition	IgM Anti-HA	HB$_s$Ag	HB$_e$Ag	Anti-HB$_e$	Anti-HB$_c$	Anti-HB$_s$
Acute HA	+	–	–	–	–	–
Early acute HB	–	+	+	–	–	–
Acute HB	–	+	+	–	+	–
Chronic HB	–	+	+/–	+/–	+	+/–
Past infection with HBV	–	–	–	–	+	+

Figure 2.1 Serologic markers of viral hepatitis.

148. Which of the following is **NOT** part of the appropriate management of infants whose mothers are HB(s)Ag positive (Figure 2.1)?
    A. Encourage the mother not to breastfeed if she develops cracked nipples
    B. Administer hepatitis B immune globulin (HBIG) as soon as possible after delivery
    C. A dose of hepatitis B vaccine should be given at the same time but in a different site as HBIG
    D. Administer a second dose of vaccine at three months of age
    E. Administer a third dose of vaccine at nine months if HB(s) Ag is negative

Questions 149–153: A 12-year-old girl presents to your office with a history of a rash for one week. The rash is annular in appearance and is slightly erythematous. There is no history of allergies, and no illness in family members. Vaccinations are up to date; the family recently returned from a vacation in the northeast. Physical exam is unremarkable, except for the rash. There has been consistent lethargy and fatigue, with only intermittent headache, fever, and chills.

149. The most likely diagnosis is
   A. atypical measles
   B. rubella
   C. tinea corporis
   D. pityriasis rosea
   E. Lyme disease

150. True statements about Lyme disease include each of the following **EXCEPT**
   A. initial signs and symptoms usually resolve without treatment in about three weeks
   B. initial lab data include a high sedimentation rate and elevated IgM
   C. the causative agent is a fungus Borrelia burgdorferi
   D. it is a tickborne illness (Ixodes genus)
   E. intermittent migrating musculoskeletal pains may be present initially

151. Which one of the following has **NOT** been described in the early phase of this disease?
   A. Encephalopathy
   B. Hepatomegaly
   C. Splenomegaly
   D. Rectal prolapse
   E. Testicular swelling

152. Late manifestations may include all of the following **EXCEPT**
   A. Bell's palsy
   B. petit mal seizures
   C. atrioventricular heart block
   D. migratory arthralgia
   E. myopericarditis

153. The treatment of choice in this patient is
   A. tetracycline
   B. keflex
   C. erythromycin
   D. penicillin
   E. topical steroid cream

Questions 154–158: A ten-year-old boy presents in your office with a history of a sore throat for one week. The family has no insurance and was waiting for the symptoms to disappear with palliative treatment (gargling and aspiring). Physical exam reveals a beefy red throat with minimal exudate, cervical adenopathy, and fever. A throat culture is positive for group A β-hemolytic strep. You immediately begin treatment with penicillin. A few weeks later he again presents to your office, complaining of fatigue and loss of appetite. He also complains of some joint pain.

Table 2.3 Jones Criteria in the Diagnosis of Rheumatic Fever

Major Manifestations	Minor Manifestations
Carditis	Fever
Polyarthritis	Arthralgia
Chorea	Previous rheumatic fever or
Erythema marginatum	rheumatic heart disease
Subcutaneous nodules	Acute phase reaction
	ESR, leucocytosis,
	C-reactive protein,
	prolonged P-R interval

154. Which of the following would establish a diagnosis of rheumatic fever using the Jones criteria (Table 2.3)?
   A. Presence of carditis
   B. Presence of carditis and arthralgia
   C. Presence of polyarthritis, fever, and evidence of recent strep infection
   D. Presence of carditis, polyarthritis, and evidence of recent strep infection
   E. Prolongation of P-R interval, fever, and arthralgia

155. True statements about rheumatic fever carditis include all of the following **EXCEPT**
   A. it occurs in 80% of initial attacks
   B. there usually are no symptoms referable to the heart
   C. the mitral valve is most commonly affected
   D. it should be suspected if there is tachycardia disproportionate to the fever
   E. the duration of carditis is usually six weeks to six months

156. The most characteristic murmur of rheumatic carditis is
- **A.** systolic ejection murmur at the left sternal border
- **B.** high-pitched late diastolic murmur
- **C.** apical high-pitched holosystolic murmur
- **D.** high-pitched early diastolic murmur
- **E.** rumbling low-pitched pan systolic murmur with radiation to the back

157. Treatment of rheumatic fever may include all of the following **EXCEPT**
- **A.** bed rest
- **B.** salicylates
- **C.** sulfadiazine
- **D.** prednisone
- **E.** oxygen

158. Which of the following is **NOT** effective for prevention of recurrence of rheumatic fever?
- **A.** Benzathine penicillin G, 1.2 million units IM every four weeks
- **B.** Oral penicillin, 200,000 units twice daily
- **C.** Tetracyclines, 500 mg qid in older patients
- **D.** Sulfadiazine, 0.5 g daily
- **E.** Sulfadiazine, 1 g daily if weight is above 30 kg

Questions 159–163: You are called to the nursery to examine an infant whose father has a history of intravenous drug abuse. The mother denies any drug use, and was discovered to be positive for HIV antibodies in her third trimester. The infant appears healthy, and his HIV antibody status is pending. His white blood count is normal, as are other baseline laboratory data. At a follow-up visit two weeks later, you inform the mother that his HIV antibody test has come back positive.

159. What percentage of infants born with maternally transferred HIV antibodies will have a persistently positive test, and go on to develop full blown AIDS?
   A. 10%
   B. 20%
   C. 40%
   D. 60%
   E. 80%

160. Clinical manifestations of AIDS in infants and young children may include all of the following **EXCEPT**
   A. hepatosplenomegaly
   B. failure to thrive
   C. chronic interstitial pneumonia
   D. chronic parotid swelling
   E. Kaposi sarcoma

161. Which of the following is **NOT** a risk factor for getting AIDS?
   A. Blood transfusion
   B. History of receiving gamma globulin
   C. Intravenous drug abuse
   D. History of receiving factor VIII concentrate
   E. Breastfeeding from an HIV-infected mother

162. Laboratory findings in AIDS may include all of the following **EXCEPT**
   A. elevated levels of IgG, IgM, and IgA
   B. altered helper/suppressor T cell ratio
   C. decrease in absolute number of T cells
   D. high levels of interferon
   E. abnormal monocyte chemotaxis

163. Which one of the following is **NOT** a common infective agent in childhood AIDS?
   A. *Bordetella pertussis*
   B. *Candida albicans*
   C. *Mycobacterium tuberculosis*
   D. *P. carinii*
   E. *Campylobacter* species

DIRECTIONS (Questions 164–188): Each group of questions below consists of lettered headings followed by a list of numbered words or statements. For each numbered word or statement, select the **one** heading that is most closely associated with it. Each lettered heading may be selected once, more than once, or not at all.

Questions 164–168:
   A. Cherry red spot of the macula
   B. Cataract
   C. Glaucoma
   D. Kayser-Fleischer ring

164. Wilson's disease

165. Lowe's syndrome

166. Galactosemia, Refsum's disease, Turner's syndrome

167. Niemann-Pick disease, mucolipidosis, Tay-Sachs

168. Cri-du-chat syndrome, Down syndrome

Questions 169–173:
   A. Achondroplasia
   B. Klippel-Feil disorder
   C. Carpenter syndrome
   D. Osgood-Schlatter disease

169. Avulsion of the infrapatellar ligament from the anterior tibial tubercle

170. Syndactyly and craniosynostosis

171. Autosomal dominant

172. Failure of segmentation of the cervical spine

173. Sprengel deformity of the scapula

Questions 174–178:
    A. Phenylketonuria
    B. Cystinuria
    C. Hyperuricemia
    D. Maple syrup urine disease

174. Detectable by newborn genetic screening, or prenatally

175. Mental retardation

176. Self-destructive biting and choreoathetosis

177. Urinary lithiasis

178. Adverse effects modified by dietary changes

Questions 179–183:
    A. XO
    B. XXY
    C. XYY
    D. XXX

179. Increased risk of neurodevelopmental abnormalities

180. Short stature, webbed neck

181. Tall stature with normal proportions

182. No characteristic phenotype

183. Gynecomastia

Questions 184–188:
    A. Pityriasis rosea
    B. Lichen planus
    C. Ichthyosis vulgaris
    D. Psoriasis

184. Genetically transmitted

185. Papulosquamous disorder

186. Hearald patch, distribution mostly on trunk, lasts for 6 to 8 weeks

187. New lesions appear on the skin at sites of trauma (Koebner's phenomenon)

188. Scaling disorder with topical treatment directed at increasing flexibility

---

DIRECTIONS (Questions 189–213): Each set of lettered headings below is followed by a list of numbered words or phrases. For each numbered word or phrase select

    A   if the item is associated with **A** only,
    B   if the item is associated with **B** only,
    C   if the item is associated with both **A** and **B**,
    D   if the item is associated with neither **A** nor **B**.

Questions 189–193:
    A. Prenatal use of alcohol
    B. Prenatal use of cocaine
    C. Both
    D. Neither

189. Increased incidence of genitourinary abnormalities

190. Mental retardation and failure to thrive are prominent features

191. Associated with congenital heart disease

192. Newborns tend to be jittery, and are easily overstimulated

193. Increased incidence of learning disorders

Questions 194–198:
    A. Gonorrhea
    B. Chlamydia
    C. Both
    D. Neither

194. May result in penile or cervical discharge

195. May cause pelvic inflammatory disease

196. Tetracyclines are the drug of choice in adolescents

197. Usual incubation period is two weeks

198. May sometimes result in lesions of the external genitalia that mimic herpes

Questions 199–203:
    A. Polyhydramnios
    B. Oligohydramnios
    C. Both
    D. Neither

199. Amniotic fluid volume between 500 and 2000 mL

200. Maternal diabetes and multiple gestations

201. Increased perinatal mortality found

202. Associated with pulmonary or renal hypoplasia

203. Found with trisomy 18,21 and anencephaly

Questions 204–208:
    A. Turner's syndrome
    B. Klinefelter syndrome
    C. Both
    D. Neither

204. May present to the pediatrician because of delayed sexual maturation

205. Normal height and proportions are found

206. Coarctation of the aorta and eye abnormalities

207. Associated with mild diabetes and chronic pulmonary disease

208. Genetic mosaics are common

Questions 209–213:
    A. Measles
    B. Rubella
    C. Both
    D. Neither

209. Single attack confers life-long immunity

210. Complications may include keratoconjunctivitis, pneumonia, and encephalitis

211. Transient polyarthritis is a common manifestation among children

212. Infection during pregnancy may result in a newborn with deafness, congenital heart disease, cataracts, and retardation

213. An enanthem may be present

DIRECTIONS (Questions 214–234): For each of the questions or incomplete statements below, **one** or **more** of the answers or completions given is correct. Select
    A if only **1, 2,** and **3** are correct,
    B if only **1** and **3** are correct,
    C if only **2** and **4** are correct,
    D if only **4** is correct,
    E if **all** are correct.

214. Which of the following statements concerning growth in the first year of life is(are) true?
  1. Birthweight doubles around five months of age and triples by one year
  2. Length increases by about 30 cm
  3. Head circumference increases by about 12 cm
  4. Most full-term infants regain their birthweight by the tenth day of life

215. Hypernatremic dehydration may be accompanied by
  1. hyperglycemia
  2. seizures
  3. high antidiuretic hormone (ADH) levels
  4. hypercalcemia

216. Which of the following drugs can be safely taken by a breastfeeding mother?
  1. Digitalis
  2. Propylthiouracil
  3. Insulin
  4. Most anticancer agents

217. Which of the following congenital malformations is(are) common in Down syndrome?
  1. Endocardial cushion defect
  2. Intestinal atresia
  3. Imperforate anus
  4. Strabismus

218. Which of the following syndromes may have blue sclera as a manifestation?
  1. Ehlers-Danlos syndrome
  2. Turner's syndrome
  3. Osteogenesis imperfecta
  4. Down syndrome

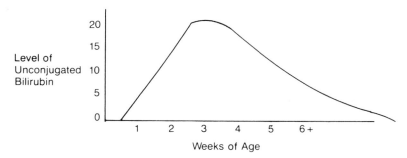

Figure 2.2 Bilirubin levels in breastfeeding jaundice.

219. Which of the following may be normal in a newborn?
   1. Phimosis
   2. Engorged breasts in a male
   3. Palpable liver, spleen, or kidneys
   4. Enlarged clitoris

220. Which of the following is(are) true of breast milk jaundice (see Figure 2.2)?
   1. Kernicterus has not been reported although levels of 20 mg% have been reached
   2. Bilirubin peaks in the second or third week
   3. It may result from a hormone in milk that inhibits the liver enzymes of the newborn
   4. Brief cessation of breastfeeding results in a decrease in the bilirubin, which increases again if breastfeeding is resumed

221. Which of the following inborn errors of metabolism is(are) amenable to dietary therapy?
   1. Galactosemia
   2. Maple syrup urine disease
   3. Homocystinemia
   4. Lesch-Nyhan disease

222. Which of the following is(are) characteristic of congenital toxoplasmosis?
1. Chorioretinitis
2. Hydrocephalus
3. Cerebral calcification
4. Microcephaly

223. Which of the following is(are) true regarding intussusception?
1. It is the most common cause of intestinal obstruction at 0 to 3 months of age
2. It may be a complication of hemolytic uremic syndrome
3. Earliest clinical manifestation is profuse bloody diarrhea, called currant jelly stool
4. Palpation of the abdomen may reveal a sausage-shaped mass in the right upper portion

224. Which of the following statements about ventricular septal defect (VSD) is(are) true?
1. It is the most common cardiac malformation
2. It commonly results in heart failure in the first week of life if it is a large defect
3. Surgical repair is not recommended for small defects
4. Pulmonary artery banding is the initial procedure used when surgery is indicated in most cases, with definitive repair at a later age

225. Stridor in the newborn may be caused by
1. congenital goiter or a vascular anomaly with tracheal compression
2. birth trauma
3. laryngomalacia
4. Pierre Robin syndrome

226. Functional or innocent murmurs in children have which of the following features?
1. They are heard in about 30% of children
2. The children have normal electrocardiograms and chest roentgenograms
3. More than one-half of normal neonates have transient systolic murmurs at the left sternal edge in the first 48 hours of life
4. A venous hum cannot be exaggerated or made to disappear by varying the position of the head or by light compression over the jugular venous system in the neck

227. Which of the following statements regarding tetralogy of Fallot is(are) true?
1. Cyanosis is always present at birth
2. Clubbing of the fingers and toes is evident by two years
3. Cardiac failure is common in the first six months of life
4. Dyspnea on exertion is a common finding

228. The EKG tracing in Figure 2.3 is characteristic of a particular dysrhythmia. Which of the following is(are) true concerning this entity?
1. It is associated with congenital heart disease in the majority of cases
2. It is treated with digoxin, which should be maintained for one year in infants
3. It is associated with Wenckebach phenomenon in the majority of cases
4. The infant looks acutely ill, has an ashen, slightly cyanotic color, and is restless and irritable

Figure 2.3

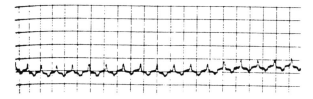

Directions Summarized				
A	**B**	**C**	**D**	**E**
1,2,3	*1,3*	*2,4*	*4*	*All* are
only	only	only	only	correct

229. Which of the following statements is(are) true of acute post-streptococcal glomerulonephritis?
   1. Usually presents abruptly with dark urine, mild edema, and decreased urinary output
   2. Early treatment of strep pharyngitis will reduce the incidence of nephritis by about half
   3. Gross hematuria usually subsides in the first week, but microscopic hematuria may persist for as long as two months
   4. Complete recovery occurs in nearly all children surviving the acute stage

230. True statements about otitis media include which of the following?
   1. Infants who develop otitis media in the first year of life have an increased risk of recurrent disease
   2. Earache and fever are not always present
   3. *Streptococcus pneumoniae* is the most common causative agent in all age groups
   4. Initial treatment of choice is still ampicillin or amoxicillin, despite an increase in the number of resistant *H. influenzae* strains

231. Which of the following have features that may make the abnormality detectable in the newborn period?
   1. XXY
   2. XO
   3. XYY
   4. Trisomy 21

232. Which of the following diseases should be screened for in the newborn period, since their outcome is improved with appropriate early management?
    1. Phenylketonuria
    2. Galactosemia
    3. Hypothyroidism
    4. Diabetes mellitus

233. Wilms' tumor is characterized by which of the following?
    1. Association with hemihypertrophy
    2. Deletion in chromosome 11
    3. Abdominal mass is the most common presenting sign
    4. The tumor is sensitive to both chemotherapy and radiotherapy

234. Which of the following is(are) true concerning iron intoxication?
    1. Nausea, vomiting, diarrhea, hematemesis, and melena occurring as a result of local necrosis begins 6 to 12 hours after ingestion
    2. Plasma iron levels are a reliable indicator of the severity of poisoning
    3. The administration of deferoxamine is unreliable as a predictor of the severity of iron poisoning
    4. Hepatic damage due to iron intoxication may begin 12 to 48 hours after ingestion

Explanatory Answers

119. B. Snuffles, or severe rhinitis, pseudoparalysis of limbs, nephrosis, and cranial bossing are all manifestations of early congenital syphilis. Other possible manifestations include failure to thrive, chorioretinitis, rash, and dactylitis. The classic *late* manifestations include saber skin (persistent periostitis of the tibia), and the Hutchinson triad of nerve deafness, interstitial keratitis, and deformed teeth. (Ref. 1, p. 644)

120. B. A positive PPD is considered to be 10 mm of induration or greater. A reaction from 5 to 9 mm is doubtful, should be repeated, and may represent exposure to atypicals. All of the other statements are true. Over 30,000 cases of tuberculosis continue to occur each year in the United States, and it must be considered in the differential diagnosis of any significant pulmonary problem. (Ref. 1, p. 631)

121. B. The child who received inactivated measles vaccine is at risk for developing atypical measles upon exposure to natural measles. This syndrome consists of high fever, pneumonia, toxicity, and a rash that may be petechial, vesicular, or urticarial, concentrated on the extremities. All children previously immunized with inactivated measles vaccine should be revaccinated with live measles vaccine to ensure their protection against natural measles. (Ref. 1, p. 655)

122. D. Neonatal herpes is usually acquired by passage through an infected birth canal. Cesarian section is recommended when a woman is culture positive, or has active lesions at term. Type II virus is usually the type responsible, and it most commonly produces a severe generalized infection that has a high mortality rate. (Ref. 1, p. 434)

123. E. Acute herpetic gingivostomatitis usually occurs between the ages of one and three years. It is characterized by fever, refusal to eat due to the multiple painful vesicles in the mouth, gingivitis, and regional lymphadenopathy. The course runs from 4 to 9 days;

avoiding dehydration is a primary therapeutic goal. Papular rash is not a feature of this disease. (Ref. 1, p. 663)

124. E. Secondary bacterial skin infection, usually due to staph or strep organisms, is the most common complication of chickenpox. Both pneumonia and encephalitis are uncommon complications of this disease. Cystitis and angioneurotic edema are not associated with varicella. About 10% of cases of Reye syndrome have an antecedent history of chickenpox. (Ref. 1, p. 666)

125. A. The rash in rubella is more variable than that of rubeola. It begins on the face, and spreads rapidly to the trunk and the rest of the body. The more prominent diagnosis feature is the retroauricular and postoccipital adenopathy. Low-grade fever, arthralgia (especially in older girls and women), and slight enlargement of the spleen are other features of this disease. (Ref. 1, p. 658)

126. B. Roseola infantum is characterized by high fever for usually three or four days, followed by a rapid return to normal temperature with the appearance of a generalized rash, which usually disappears within 24 hours. The disease is most common between the ages of 6 and 18 months, and the etiology is unknown. (Ref. 1, p. 661)

127. D. In an infant with pyloric stenosis, palpation of the abdomen may reveal a hard, mobile, nontender, olive-shaped mass under the edge of the liver. This finding, coupled with an appropriate history, provides a diagnosis and makes a barium swallow unnecessary. All of the other question choices are characteristics of an infant with pyloric stenosis. (Ref. 1, p. 778)

128. E. There is no evidence that circumcision protects against penile cancer, which is a rare disease in this country. Since the 1970s the American Academy of Pediatrics has stated that there is no medical indication for routine circumcision. (Ref. 1, p. 1163)

129. E. All of the statements are true. Hodgkin's disease has a bimodal frequency, with peak incidence rates in the 15-to-34 and over-50 years age groups. It is more common in boys than girls. Enlarged cervical lymph nodes, which are firm, nontender, discrete, and single or multiple, is the most common presentation. There are usually few systemic manifestations initially. Recent advances in therapy have resulted in a high cure rate. (Ref. 1, pp. 1089–1091)

130. B. Orthostatic proteinuria is frequently discovered on a routine urinalysis. It can be present in children with renal disease, but most patients are healthy without any underlying pathology, nor with any future predisposition to the development of hypertension. The diagnosis is confirmed by the absence of other abnormalities in the urine, and the demonstration that the total 24-hour urinary excretion of protein is less than 1000 mg, as well as the fact that the first morning urine specimen is relatively free of protein, and that there is a significant increase in the excretion of protein when the child is upright. The reference book contains all the details of this test. (Ref. 1, p. 1128)

131. B. Wilms' tumor, polycystic kidneys, subacute bacterial endocarditis, and renal artery thrombosis may all present with a variable degree of hematuria. Minimal lesion nephrotic syndrome presents with edema and proteinuria. (Ref. 1, p. 1114)

132. E. Body size and level of sexual maturation of the patient are the most important factors in determining insulin dosage. There are, however, a number of factors that can alter the fine tuning of the dosage requirement. These include acute infections, stress, the onset of puberty, and the development of insulin antibodies. There is no association of diabetes with hypothyroidism. (Ref. 1, pp. 1256–1258)

133. C. Lymphocytic thyroiditis is more common in girls, reaches a peak incidence in adolescence, and most commonly results in a diffusely enlarged, firm, nontender goiter, which appears insidiously. The clinical course is variable and not all cases progress to hypothyroidism. (Ref. 1, p. 1198)

134. D. There is no evidence of increased susceptibility to staph infections in this disease. There is an increase in salmonella osteomyelitis. Because of the functional asplenia, and a serum opsonin deficiency, these patients are also susceptible to pneumococcal septicemia and meningitis. Pneumococcal vaccine is recommended. The disease may be found in Hispanics as well as blacks, due to intermarriage between the groups. It usually presents in the latter part of the first year of life, with anemia or painful crisis. The disease usually results in a hemoglobin concentration of 6 to 8 g/dL. Since there is no definitive treatment, therapy is indicated only during acute episodes. (Ref. 1, p. 1050)

135. C. Genu valgum (knock knees) is the only condition mentioned which does not result in an intoeing gait. Progressive knock knees after age 6 should be evaluated for any underlying abnormality. Tibial torsion and increased anteversion of the hip usually improve with age, and are more cosmetic than functional problems in most cases. Metatarsus adductus should be corrected in early infancy with serial casting if it is severe. (Ref. 1, p. 1344)

136. E. Leukocoria is the term given to a white pupillary reflex. Examination under anesthesia is usually necessary to establish the diagnosis. The differential diagnosis includes retrolental fibroplasia, retinoblastoma, cataract, toxocara cyst, retinal dysplasia, exudative retinopathy, fundus coloboma, and persistent hyperplastic primary vitreous, but *not* iridocyclitis. (Ref. 1, p. 1452)

137. C. Use of the Glasgow Coma Scale has become an important component of evaluating neurologic conditions in children. This is especially true when a child has experienced any degree of head trauma. To use this scale, elicit the best response in each of the three areas. If you are unable to test any of the areas, assign it no score. A total of 7 or less indicates severe compromise of brain function. (Ref. 1, p. 1284)

138. D. Acute leukemia is the most common site of malignancy in the under-five-years-old age group, representing 36% of all cancers. After age 10, it drops to only 18% of cancers and by age 15 falls to 12%. Lymphoma and brain tumors become the more common malignancies as the child ages. (Ref. 1, p. 1082)

139. D. The case presented is that of a child with Reye syndrome,
140. B. which is associated with outbreaks of influenza, especially
141. D. B, varicella, and use of aspirin. It is important to have a
142. B. high index of suspicion for the disease. Diagnosis can be
143. C. made using the staging system, and an abnormal liver profile, prothrombin time, and serum ammonia level will be found

early in the process. A liver biopsy is helpful, but not essential. Early support, including administration of vitamin K, monitoring of intracranial pressure (and treatment if elevated), and avoidance of hypoglycemia, can be lifesaving. Prognosis is most related to the duration of disordered cerebral function during the acute stage. (Ref. 1, pp. 840–842)

144. A. Use of drugs during pregnancy continues to be a national
145. A. problem. The focus of this discussion is on a mother who,
146. D. through IV drug abuse, is a carrier of hepatitis B surface
147. B. antigen. The CDC, American Academy of Pediatrics, and
148. A. American College of Obstetricians and Gynecologists have all recommended *universal* screening of pregnant women. Mothers who are hepatitis B "E" antigen positive are at greater risk of transmission to their offspring. Transmission probably occurs at the time of delivery, since only 2.5% of infants are antigen positive at birth. Management of the infant includes administration of hepatitis B immune globulin (HBIG) and the first of three doses of vaccine at birth. The other doses are given at three months and nine months. If the infant is antigenemic at nine months, he is considered a vaccine failure and is managed as a carrier. Breastfeeding is *not* contraindicated. (Ref. 1, pp. 685–688)

149. E. Lyme disease is a tickborne spirochetal disease caused by
150. C. Borrelia burgdorferi. It is characterized by a distinctive
151. D. skin lesion known as erythema chronicum migrans. Initial
152. B. signs and symptoms often resolve within 3 to 4 weeks,
153. A. regardless of treatment, though the rash may recur. There is then a latent period of weeks to months, after which time neurologic, cardiac, or musculoskeletal manifestations may appear. The classic neurological triad is meningitis, cranial neuropathy (including Bell's palsy) and peripheral radiculoneurapathy. Cardiac involvement may include AV block pericarditis of cardiomegaly. Migratory arthralgias and arthritis involving large joints, especially the knee, are the most common latent manifestation. Treatment of choice is penicillin in younger children and tetracycline in older children. (Ref. 1, p. 533)

154. D. Rheumatic fever is experiencing a resurgence in this
155. A. country. Numerous states have reported an increase in
156. C. cases. The diagnosis is best made using the Jones criteria.
157. C. Two major, or one major plus two minor, criteria, in
158. C. addition to evidence of recent strep infection, indicate a
high probability of rheumatic fever. Carditis occurs in 50% of initial
attacks, with the onset of a characteristic apical, high-pitched,
blowing, holosystolic murmur of mitral valvulitis. Treatment of
carditis involves bed rest and salicylates. If carditis and cardiomegaly
are present, prednisone should be given. Mild heart failure should be
managed with complete bed rest, oxygen, fluid restriction, and
steroids. Prophylaxis against strep is recommended at least throughout adolescence (life-long if valvular disease is present), using an
appropriate regimen. (Ref. 1, pp. 539–543)

159. C. Pediatricians are being faced daily with the reality of caring
160. E. for infants with AIDS. The fastest growing segment of
161. B. patients with pediatric AIDS is described in this patient.
162. D. Perinatal transmission from a drug abusing HIV positive
163. A. mother accounts for the greatest number of children with
AIDS. Of infants who test positive at birth, 30% to 50% will go on
to develop full blown AIDS. Receiving gamma globulin is not a risk
factor for AIDS. In contrast to adult AIDS, Kaposi sarcoma is rare in
children. The list of opportunistic infections to which both children
and adults are susceptible is long, but does not include pertussis.
Mortality is about 80% within three years. (Ref. 1, pp. 467–469)

164. D. It is important to be able to recognize certain opthalmic
165. B,C. entities and to realize that they often do not exist in
166. B. isolation, but herald a larger syndrome. Some defects,
167. A. such as cataracts, are found in a number of diseases, such
168. B. as Lowe's syndrome, galactosemia, Turner's syndrome,
Refsum's disease, Down syndrome and Cri-du-chat. The cherry red
spot of the macula is more specific to storage diseases, as the
Kayser-Fleischer ring is to Wilson's disease. (Ref. 2, pp. 1762–1763)

169. D. Achondroplasia is the most common genetic skeletal
170. C. dysplasia. Its transmission is autosomal dominant and is
171. A. easily recognized in the newborn period by the short
172. B. stature, large head, and depressed nasal bridge. Klippel-
173. B. Feil syndrome involves a short neck, limited neck motion,
and a low occipital hairline. There are varying degrees of fusion of
the cervical vertebrae. It is often associated with Sprengel's
deformity, an abnormally high scapula. Carpenter syndrome is
transmitted by autosomal recessive inheritance, and can be diagnosed
prenatally. Osgood-Schlatter disease presents with knee pain, with
swelling or tenderness over the tibial tubercle. It is self-limiting, and
disappears as growth of the proximal tibia ceases. (Ref. 2, pp. 334,
354, 1576, 1804, 1821)

174. A,C,D. In phenylketonuria, phenylalanine cannot be con-
175. A,C,D. verted to tyrosine, and accumulates in the brain,
176. C. interfering with the production of serotonin and
177. B. aminobutyric acid. Cystinuria is a defect in renal
178. A,B,D. tubular function in which the reabsorption of cystine,
lysine, arginine, and ornithine is limited. Because cystine is
insoluble, renal calculi are formed at an early age. Hyperuricemia is
a disorder of purine metabolism, resulting in severe mental
retardation, cerebral palsy, choreoathetosis and self-biting. It is
transmitted as an X-linked characteristic. Maple syrup urine disease
is a disease characterized by high blood and urine concentrations of
leucine, isoleucine, and valine. Symptoms often begin in the first
week, and progress rapidly to death within 2 to 4 weeks. It is an
autosomal recessive disease, and dietary treatment, though difficult,
can be useful when the diagnosis is made early. (Ref. 2, pp. 238, 239,
243, 255)

179. B,D. XO karyotype is that of Turner's syndrome. It is present
180. A. in about 1/2500 female newborns, and is characterized
181. C. by short stature, gonadal dysgenesis, and a variety of
182. C, D. somatic findings, including webbed neck, and widely
183. B. spaced nipples. XXY karyotype is that of Klinefelter
syndrome, with an incidence of about 1/1100 male newborns. It is
characterized by small testes, gynecomastia, sterility, and neurodevelopmental abnormalities. XYY occurs in about 1/1000 male
newborns, and is characterized by tall stature. Impulsiveness and

criminal behavior have not been supported by recent studies. XXX occurs in 1/1200 female newborns and also has an increased risk of neurodevelopmental abnormalities, with an increased incidence of subnormal intelligence. (Ref. 2, p. 238)

184. B,C. Pityriasis rosea occurs primarily on the trunk. Occa-
185. A,B,C. sionally there is a prodrome of fever and mild sore
186. A. throat, suggesting a viral origin. It is often misdiag-
187. B,D. nosed as tinea corporis. Treatment is directed at
188. C. reducing itching. Lichen planus is also a papu-
losquamous disease that shares Koebner's phenomenon with psoriasis. The buccal mucosa is often involved, and the cause is unknown. There is some HLA association. Ichthyosis vulgaris has an incidence of 1/1000, and has autosomal dominant inheritance. Fine, branny scales on the trunk and extensor surfaces of the extremities are characteristics. The flexor surfaces are spared. The aim of topical treatment is to increase flexibility by hydration. The typical lesions of psoriasis are red plaques with a silvery-white scale, especially in the elbows, knees, scalp, and penis. Topical coal tar products are superior to topical steroids for treatment. Use of UV light or methotrexate are extremely limited in children. (Ref. 2, pp. 794–798)

189. B. The recent explosion of crack use nationwide has also
190. A. involved pregnant women. An estimated 8 million Ameri-
191. A. cans use cocaine regularly, and most are between 20 and 35
192. C. years, the prime childbearing years. Cocaine babies
193. C. experience an increase in GU anomalies, including hypospadias, hydronephrosis, and prune belly syndrome. They also tend to be jittery and easily overstimulated as newborns. Long-term studies are in progress, and early results suggest an increase in learning disabilities. The difficulty is that most users are polydrug users, making it less clear which drugs cause what clinical findings. Fetal alcohol syndrome has been well described, and may include cardiac septal defects, mental retardation, joint and limb abnormalities, short palpebral fissures, epicanthal folds, micrognathia, and small for gestational age birth weight. (Ref. 2, pp. 373, 752)

194. C. Nongonoccal urethritis (NGU) is now the most common
195. C. sexually transmitted disease. Most cases are caused by
196. B. chlamydia, with an incubation period of about two weeks,
197. B. in contrast to 3 to 5 days with gonorrhea. In the male there
198. D. is penile discharge and/or pain on urination, though usually
less dramatic than that found with gonorrhea. Females are most often
asymptomatic, though as with gonorrhea, they may complain of
discharge, or of abdominal pain or pain on lovemaking if the
infection has produced a PID syndrome. Tetracycline 500 mg qid is
the drug of choice. Neither mimics herpes in any way. (Ref. 2, pp.
493, 501)

199. D. Polyhydramnios is an estimated third trimester amniotic
200. A. fluid volume greater than 2000 mL; oligohydramnios has a
201. C. volume less than 500 mL. Both are associated with
202. B. conditions that increase perinatal mortality. Polyhydram-
203. A. nios is found with maternal diabetes, multiple gestations,
anencephaly, trisomy 18 or 21, atresias of the upper intestinal tract,
and erythroblastosis fetalis. Oligohydramnios is associated with
hypoplasia or aplasia of fetal kidneys, pulmonary hypoplasia, and
abnormalities of the fetal limbs. (Ref. 1, p. 366)

204. A. Turner's syndrome (XO) is characterized by short stature,
205. D. streak gonads, sexual infantilism (often the presenting
206. A. complaint), and somatic anomalies including broad chest,
207. B. webbed neck (often detectable in the newborn, along with
208. C. lymphedema), and short fourth metacarpal. Coarctation of
the aorta and eye abnormalities such as cataracts, blue sclera, and
strabismus may also be found. Klinefelter syndrome (XXY) is
characterized by small testes, sterility, gynecomastia, disproportionately long legs, and subnormal intelligence. Diabetes, pulmonary
disease, and varicose veins occur with increased frequency. Both
syndromes may have mosaics. (Ref. 2, pp. 1536–1538)

209. C. A single attack of either rubella or measles confers
210. A. life-long immunity. Both diseases are vaccine preventable.
211. D. Both may have an enanthem which appears before the rash.
212. B. In measles, Koplik's spots appear on the buccal mucosa. In
213. C. rubella, there may be punctate or larger red spots on the soft
palate. Rare complications of rubella include encephalitis and

thrombocytopenic purpura. Polyarthritis is more common in women, and is uncommon in children. The congenital rubella syndrome is where prevention efforts are placed. Measles complications include keratoconjunctivitis, myocarditis, pneumonia, secondary bacterial infections, encephalitis, and subacute sclerosing panencephalitis. (Ref. 2, pp. 596, 606)

214. E. Birthweight does double by about five months of age, and triple by about one year. Head circumference does increase by about 12 cm in the first year of life, the average one-year-old having a head circumference of about 45 to 48 cm. During this first year, the average infant will grow 25 to 30 cm, or 10 to 12 inches. Most full-term infants regain their birthweight by 10 days of age, and gain about 20 g/d for the first five months. (Ref. 1, p. 27)

215. A. The main principal in the treatment of hypernatremic dehydration is to return the serum sodium to normal levels slowly. This allows equilibration of fluid and electrolytes within the brain, and avoids seizures. Hyperglycemia may be present but usually declines with rehydration. High ADH levels reflect the body's attempt to retain fluids in the face of a depleted vascular volume. *Hypo*calcemia may develop during rehydration, and can be avoided by giving appropriate amounts of potassium in the replacement therapy. (Ref. 1, p. 194)

216. A. The same caution used during pregnancy concerning drugs should be extended to nursing mothers. Almost all drugs will enter into breast milk, though many do so in small amounts with no adverse effects on the infant. Drugs that are definitely contraindicated in nursing mothers include lithium and phenindione, with temporary cessation of nursing recommended when taking metronidazole or certain radiopharmaceuticals. (Ref. 1, p. 364)

217. E. All of the malformations listed may be seen in Down syndrome. In addition, one can find mental retardation, hypotonia, speckled irides (Brushfield spots), protruding tongue, malformed ears, cryptorchidism, simian palmar creases, furrowed tongue, short broad hands with hypoplasia of the middle phalanx of the fifth finger, and a gap between the first and second toes. (Ref. 1, p. 255)

218. A. Of those syndromes mentioned, only Down syndrome does not exhibit blue sclera. All of the syndromes may have additional eye abnormalities, including acrus senilis and corneal deformities (osteogenesis imperfecta), cataracts and color blindness (Turner's), keratoconus, ectopia lentis, and retinal detachment (Ehlers-Danlos). (Ref. 2, pp. 1762–1763)

219. A. Phimosis is the normal condition of the newborn penis, with gradual loosening of the foreskin in the first few years of life. Engorged breasts may be present normally in the first few days of life in either sex, due to transplacental passage of maternal hormones. The liver, spleen, and kidneys may be palpable in the newborn period, though they are normally not enlarged. An enlarged clitoris is not normal, and is most likely secondary to congenital adrenal hyperplasia. (Ref. 1, p. 7)

220. A. The cause of breast milk jaundice remains unknown. Recent attention has focused on the possible role of the enterohepatic pathways, and the concentration of free fatty acids in the breast milk, although a hormonal etiology has not been ruled out. Kernicterus has never been reported. The jaundice usually peaks in the second or third week. Breastfeeding is *not* a cause of increased jaundice in the first few days of life! Cessation of breastfeeding for 24 to 48 hours usually results in a drop in the bilirubin level, which then tends to remain constant when breastfeeding is resumed. (Ref. 1, p. 407)

221. A. Galactosemia and maple syrup urine disease result in mental retardation if left untreated; both are amenable to dietary therapy, as is homocystinemia, a defect in the metabolism of methionine which can result in ectopia lentis, osteoporosis, and thromboembolic episodes. Other inborn errors amenable to dietary therapy include phenylketonuria and propionicacidemia. Lesch-Nyhan disease is an X-linked disorder of purine metabolism that currently has no treatment. (Ref. 1, pp. 285, 289, 307)

222. E. Chorioretinitis, hydrocephalus, cerebral calcification, and microcephaly may all be components of congenital toxoplasmosis. Additional characteristics may include psychomotor retardation, seizure disorders, and ocular palsies. Maternal infections characteristically are often asymptomatic, and their offspring are often not infected at all. (Ref. 1, p. 734)

223. D. Intussusception is rare under three months of age. It is the most common cause of intestinal obstruction between two months and six years of age. It is not associated with the hemolytic uremic syndrome. It presents with the sudden onset of severe paroxysmal pain in a previously well infant, who appears well and may even play between episodes. Currant jelly stool is a later manifestation that is not found in all cases. Careful palpation of the abdomen may reveal a mass in the right upper portion, which is painful and somewhat ill defined. (Ref. 1, p. 787)

224. B. Ventricular septal defects will produce symptoms according to the size of the defect and the pulmonary blood flow and pressure. There is usually a harsh blowing pansystolic murmur, heard best over the lower left sternal border, which may not be heard in the first days of life. Infective endocarditis is a possible complication, and appropriate antibiotic prophylaxis is needed during dental work, oropharyngeal surgery, and GU instrumentation. Surgical repair, when indicated, is most often done in a one-step definitive operation. (Ref. 1, p. 981)

225. E. Congenital goiter, vascular anomalies, birth trauma, laryngomalacia, and Pierre Robin syndrome (micrognathia with pseudomacroglossia, glossoptosis, and high arched or cleft palate) can all cause stridor in the newborn. Additional causes include tumors of the larynx, branchial cleft cysts, hemangioma, and lymphangioma. (Ref. 1, p. 855)

226. A. Once the clinician has become familiar with the most common innocent murmurs, diagnostic workup is unnecessary. Chest x-ray and EKG are normal, and the quality, location, and variability indicate their innocent nature. More than half of normal neonates in the first 48 hours, and 30% of children, may demonstrate these murmurs. A characteristic of a venous hum is that it *can* be exaggerated or made to disappear by the procedures described. (Ref. 1, p. 946)

227. C. Tetralogy of Fallot consists of ventricular septal defect, pulmonary stenosis, right ventricular hypertrophy, and dextroposition of the aorta. Cyanosis may not be present at birth, though it usually appears later, and clubbing is evident by age two years. The severity of the lesion is often indicated by the degree of cyanosis. Cardiac failure is uncommon, but may occur when there is severe iron deficiency. Dyspnea on exertion is common, and children will assume a squatting position in order to relieve it. (Ref. 1, pp. 964–968)

228. C. Paroxysmal atrial tachycardia in infancy frequently results in heart failure manifested by an ill, irritable, slightly cyanotic infant with tachypnea and hepatomegaly. It is not associated with congenital heart disease or the Wenckebach phenomenon. It may be a component of the Wolff-Parkinson-White syndrome (short pr interval, D wave). Vagal stimulation may abolish the attack, but affected infants should still be treated with digoxin, since the recurrence rate is high. (Ref. 1, p. 1006)

229. E. Recent data seems to indicate that the incidence of nephritis can be reduced by the prompt treatment of streptococcal pharyngitis, although the effect of early treatment of impetigo is unknown. Edema, dark-colored urine, decreased urine output, abdominal pain, general malaise, and a low-grade fever are the usual manifestations. Cardiac, hypertensive encephalopathy, and acute renal failure are dreaded complications. Gross hematuria resolves within a week, though microscopic cells and casts frequently persist for up to two months. The long-term prognosis is excellent. (Ref. 1, p. 1119)

230. E. Prophylactic antibiotics, in the form of a daily dose of sulfonamides or amoxicillin, can be useful in those children with recurrent otitis media. Normally, children will improve clinically within 48 hours after antimicrobial therapy. If there is no improvement, the possibility of a resistant organism must be suspected, and trimethoprimsulfamethoxazole or erythromycin and a sulfonamide may be given. (Ref. 1, pp. 879–883)

231. C. XXY and XYY have no features which distinguish them in the newborn period. Turner's (XO) may show lymphedema and webbed neck. Trisomy 21 shows the characteristic phenotype of Down syndrome. (Ref. 2, p. 226)

232. A. Patients with phenylketonuria and galactosemia benefit markedly from early dietary therapy. Hypothyroidism also needs to be detected early so that replacement therapy can begin as soon as possible to minimize retardation. Screening for hypoglycemia in at risk infants makes excellent sense. There is no need to screen for diabetes in early infancy, regardless of family history. (Ref. 2, p. 226)

233. E. Wilms' tumor is associated with hemihypertrophy, GU abnormalities, and sporadic aniridia. Most Wilms' tumors submicroscopically demonstrate a deletion in chromosome 11. It is a solitary growth that arises in any part of either kidney, and most often presents as an asymptomatic abdominal mass. Surgical removal of the involved kidney is the first step, which is usually followed by chemotherapy and radiation, if beyond stage I. (Ref. 1, pp. 1095–1097)

234. D. Iron poisoning can be thought of in four phases. (1) Local irritative effects: the hematemesis, melena, diarrhea, and abdominal pain usually *subside* within 6 to 12 hours. (2) Stage of quiescence: because of the rapid transport of iron into the liver in the first two phases, plasma iron levels are unreliable as indicators of the severity of the poisoning. (3) This phase begins 12 to 48 hours after ingestion and is due to the toxic effects on the liver. This is when most deaths occur. (4) There may, if recovery has occurred, be eventual stenosis of the pyloric area due to the severe irritation of the first phase. A single dose of deferoxamine can be used to assess the severity of the poisoning. Deferoxamine will bind excess iron, and cause the urine to become pink or red. (Ref. 1, p. 1500)

References

1. Vaughan VA III, McKay RJ et al (eds): *Nelsons Textbook of Pediatrics,* ed 13. Philadelphia, WB Saunders Co, 1987.

2. Rudolph AM et al (eds): *Pediatrics,* ed 18. Norwalk, CT, Appleton & Lange, 1987.

3 Preventive Medicine and Public Health

Faculty Members with
Richard H. Hart, MD, Chairman
Department of Public Health and
Preventive Medicine
Loma Linda School of Medicine

DIRECTIONS (Questions 235–294): Each of the questions or incomplete statements below is followed by five suggested answers or completions. Select the **one** that is best in each case.

---

235. Which of the following is unrelated to the chi-square test of statistical significance?
    A. Qualitative data
    B. Degrees of freedom
    C. Null hypothesis
    D. Significance level
    E. Life table

236. Criteria for using a z test include all of the following **EXCEPT**
    A. an r value (correlation coefficient)
    B. random samples
    C. quantitative data
    D. normal distribution
    E. population standard deviation known

237. Which of the following is **NOT** related to the performance of a significance test?
    A. Calculate a p value
    B. Accept a null hypothesis
    C. Reject a null hypothesis
    D. Draw specific conclusions
    E. Perform an indirect standardization

238. Advantages of prospective studies include all of the following **EXCEPT**
    A. lack of bias in factor
    B. relatively inexpensive
    C. can yield incidence rates as well as relative risk
    D. can yield associations with additional disease as a by-product
    E. efficient for studying rare exposure

239. Advantages of retrospective studies include all of the following **EXCEPT**
    A. rather inexpensive to carry out
    B. number of subjects can often be rather small
    C. results can be obtained rather quickly
    D. yield rates rather than ratios
    E. frequently identify more than one risk

240. In August 1984, three cases of botulism were reported to the Centers for Disease Control. The ill persons had eaten improperly handled food. Which of the following statements is true concerning botulism?
    A. The spores of *Clostridium botulinum* are ubiquitous in soil
    B. The spores are heat labile
    C. Food-borne botulism in the United States generally results from contaminated meat
    D. The botulinum toxin is heat resistant
    E. Spores germinate and produce toxins without regard to the pH and oxygen content of the food

241. The provision of a safe public water supply involves all of the following **EXCEPT**
  A. primary sedimentation and clarification by exposure to air and light
  B. removal of bacteria and algae by the addition of a coagulant
  C. passage through a sand filter
  D. the addition of a disinfectant, usually fluorine in a concentration of 1.5 to 3 parts per million
  E. prevention of contamination of purified water by prevention of cross connections and back syphonage

242. Regarding refuse disposal in the United States, which statement is **INCORRECT**?
  A. Most disposal is by simple open dumping
  B. Salvage and composting together accounts for about 50% of refuse disposal
  C. Most urban communities are rapidly running out of places to dump or bury
  D. About two-thirds of the cost of solid waste disposal is represented by pick-up and transfer
  E. The United States produces about 3.5 billion tons of solid waste annually

243. Sulfate of aluminum is commonly utilized in
  A. milk purification
  B. decreasing air pollution
  C. prevention of radiation hazard
  D. sewage disposal
  E. water purification

244. Digestion, settling, trickling, and drying are all associated with
  A. water purification
  B. sludge treatment
  C. air pollution control
  D. milk processing
  E. garbage disposal

245. Which of the following is perhaps the most prevalent form of malnutrition in the United States and other industrialized countries?
  - **A.** Hypervitaminosis
  - **B.** Hypovitaminosis
  - **C.** Pellagra
  - **D.** Anorexia nervosa
  - **E.** Obesity

246. The most frequently eaten raw shellfish most often involved in disease outbreaks, including typhoid fever and viral hepatitis, is
  - **A.** oysters
  - **B.** crabs
  - **C.** clams
  - **D.** lobsters
  - **E.** scallops

247. Which of the following is characteristic of tetanus and its transmission?
  - **A.** It is due to nonspore-bearing aerobic bacillus
  - **B.** It is found in the intestinal tract of many animals, including humans
  - **C.** Early and thorough cleansing of wounds is now considered unimportant
  - **D.** Antibiotics and disinfectant have been proven to be of value in prevention and therapy
  - **E.** Tetanus toxoid is one of the least effective of all immunologic agents

248. Illnesses caused by adenoviruses include all of the following **EXCEPT**
  - **A.** common respiratory disease
  - **B.** epidemic keratoconjunctival
  - **C.** pharyngoconjunctival fever
  - **D.** "fever blisters" (herpes simplex)
  - **E.** pneumonia

Table 3.1 Water-borne Disease Outbreaks, by Year and Type of System, United States, 1971–1980

	Community	Noncommunity	Private	Total	Total cases
1971	5	10	4	19	5,182
1972	10	18	2	30	1,650
1973	5	16	3	24	1,784
1974	11	10	5	26	8,363
1975	6	16	2	24	10,879
1976	0	23	3	35	5,068
1977	12	10	3	34	3,860
1978	10	19	4	32	11,435
1979	23	14	4	41	9,720
1980	23	22	5	50	20,008
Total (%)	(36)	(53)	(11)		

Reproduced from "Water-related Disease Outbreaks Surveillance" Annual Summary 1980, from the U.S. Department of Health and Human Services, Public Health Service.

249. In Table 3.1, the 20,008 cases in 1980 represent
- **A.** incidence data
- **B.** point prevalence data
- **C.** period prevalence data
- **D.** gross overreporting
- **E.** crude mortality rates

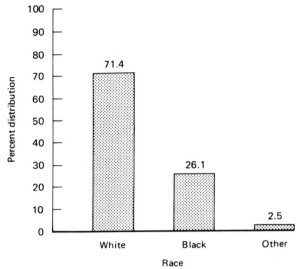

Figure 3.1 Percent distribution of female family planning patients by race: United States, 1980. Reproduced from NCHS Advance Data, Number 82, June 16, 1982, from Vital and Health Statistics of the National Center for Health Statistics, U.S. Department of Health and Human Services.

250. The authors, in referring to Figure 3.1, state that it "shows that the majority of women who visited a family planning clinic were white women." With regard to that statement,
   A. the authors are correct
   B. this represents the "fallacy of comparing percentages"
   C. this represents the "fallacy of the average"
   D. conclusions based on the data are of little value because there are no controls
   E. the data as presented represents a "normal" distribution

251. Regarding recommendations for tetanus prophylaxis in routine wound management, which of the following is false?

A. TIG (tetanus immune globulin) is recommended for those with dirty wounds who have unknown or <3 doses of tetanus toxoid

B. TIG is recommended for those with dirty wounds who have ≥3 doses of tetanus toxoid

C. TD (tetanus toxoid) is recommended for those with clean minor wounds who have had unknown or <3 doses of tetanus toxoid, with the last dose having been within the last ten years

D. TD is recommended for those with dirty wounds who have had unknown or <3 doses of tetanus toxoid

E. TD is recommended for all wounds

252. For which of the following vaccines is the recommended route of administration intramuscular?

A. BCG

B. Measles

C. MMR

D. OPV

E. Tetanus

253. Which of the following statements regarding antigens is correct?

A. Live antigens may be given several days apart

B. There should be an interval of at least one week between the administration of two live antigens

C. There should be an interval of at least four weeks between the administration of two live antigens, if not administered simultaneously

D. A live and killed antigen should not be given together

E. Killed antigen vaccines are better than live antigen vaccines

254. Which one of the following should be considered a routine contraindication to vaccination?
- **A.** Mild acute illness with low-grade fever or mild diarrheal illness in an otherwise well child
- **B.** Current antimicrobial therapy of the convalescent phase of an illness
- **C.** Pregnancy in a household member
- **D.** Recent exposure to an infectious disease
- **E.** Moderate to severe febrile illness

255. In general, immunocompromised hosts may safely receive all the following vaccines **EXCEPT**
- **A.** influenza
- **B.** pneumococcal
- **C.** hepatitis B in double doses
- **D.** MMR
- **E.** tetanus toxoid

256. The recommended schedule for active immunization of normal infants and children is which of the following?
- **A.** DTP at 2, 4, 6, 8, and 10 months
- **B.** DTP at 2, 4, 6, and 15 months plus booster at 4 to 6 years
- **C.** MMR before 12 months
- **D.** MMR at 18 years
- **E.** TD every five years

257. After having received recombinant DNA hepatitis B vaccine, which of the following represents an immune response?
- **A.** HBsAG
- **B.** Anti-HBs
- **C.** HBcAG
- **D.** IGM Anti-HBc
- **E.** All indicate an immune response to the vaccine

258. Current concepts/guidelines for measles immunization include all of the following **EXCEPT**

  A. two doses: the first at 12 to 15 months; the second at age 4 to 6 years

  B. physician diagnosed measles is acceptable for determining positive immune status

  C. laboratory evidence of past measles is a way for determining immunity

  D. exposure to measles is a contraindication to vaccination

  E. some persons who received inactivated vaccine are at risk of developing a typical measles syndrome when exposed to natural virus

259. The two most important values usually necessary as a description of the frequency distribution of a series of observations are

  A. standard deviation and mean

  B. median and variance

  C. mode and range

  D. range and mean

  E. size of sample and standard deviation

260. For measurements that follow a normal distribution, the values that will differ from the mean by more than three times the standard deviation are approximately.

  A. 1 in 10

  B. 1 in 25

  C. 1 in 55

  D. 1 in 100

  E. 1 in 370

261. Assuming a normal curve, the proportion of observations that lie within three times the standard deviation from the mean is

  A. 99.73%

  B. 95.45%

  C. 12.20%

  D. 68.27%

  E. 00.27%

Figure 3.2

262. The graph in Figure 3.2 is called a
   A. histogram
   B. chi-square
   C. scatter diagram
   D. bell curve
   E. spot map

263. In Figure 3.3, what percentage of values would you expect to find within the shaded area?
   A. 68%
   B. 95%
   C. 34%
   D. 75%
   E. Some other value

Figure 3.3

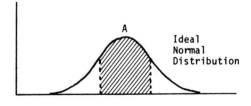

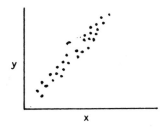

Figure 3.4

264. The slope of the graph shown in Figure 3.4 represents a
   A. positive linear correlation
   B. negative linear correlation
   C. curvilinear correlation
   D. quadratic correlation
   E. zero correlation

265. In Figure 3.5, point A would be best described by all of the following **EXCEPT**
   A. mean
   B. median
   C. mode
   D. x
   E. range

Figure 3.5

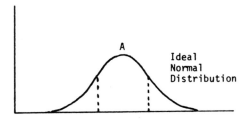

266. Schistosomiasis, or bilharziasis, is an ancient helminthic infection of the mesenteric, portal, and pelvic venous systems. At least 150 million people are currently affected. Effective control has been rare. The vector involved is
    A. man
    B. fish
    C. snails
    D. mosquitoes
    E. cercariae

267. The cholera vibrio proliferates within the lumen of the intestine and
    A. invades the bloodstream
    B. invades the tissues
    C. inhibits sodium transport
    D. produces little effect on electrolytes
    E. allows resorption of isotonic fluids

268. The virus most often associated with colds in adults is
    A. rhinovirus
    B. coxsackie virus
    C. ECHO virus
    D. respiratory syncytial (RSV) virus
    E. parainfluenza virus

269. Influenza is characterized epidemiologically by all of the following EXCEPT
    A. epidemic-pandemic potentiality
    B. excess mortality among predictable high-risk groups
    C. high morbidity–low mortality
    D. periodic-cycle recurrences
    E. in temperate zones, epidemics tend to occur in the late summer

270. The most effective means of temporary contraception that a woman can use is
    A. intrauterine devices
    B. oral contraception (the pill)
    C. spermatocidal jelly
    D. diaphragm
    E. condom

271. Of the following, which is **NOT** a contraindication for using oral contraceptives?
    A. Coronary artery disease
    B. Stroke
    C. breastfeeding an infant over six weeks of age
    D. Pregnancy
    E. Impaired liver function

272. The tendency for the number of births in a population to continue increasing despite a fertility decline is referred to as
    A. crude birth rate
    B. population momentum
    C. fecundity phenomenon
    D. fertility gap
    E. population retardation

273. Recent U.S. data concerning teenage pregnancy and fertility reveals which of the following to be true?
    A. Decreased proportions of unmarried teens having sexual intercourse
    B. Decrease in premarital conceptions
    C. Decreased contraceptives use
    D. Increased abortion among first premarital pregnancies
    E. Increased legitimization of births through marriage

274. Fluoridation of water is a method of
    A. decreasing bacterial population
    B. preventing growth of bacteria
    C. preventing bacterial toxin formation
    D. limiting multiplication of viruses
    E. preventing dental decay

275. The leading cause of maternal deaths in the United States is
    A. toxemia
    B. hemorrhage
    C. sepsis
    D. respiratory complications
    E. cardiovascular complications

276. All of the following disorders are believed to be solely the result of genetic factors **EXCEPT**
    A. epilepsy
    B. hemophilia
    C. fibrocystic disease of the pancreas
    D. multiple polyposis of the colon
    E. xeroderma pigmentosum

277. Which of the following is the most expensive occupational health problem?
    A. Circulatory diseases
    B. Bone and joint disorders
    C. Digestive disorders
    D. Allergic disorders
    E. Upper respiratory infections

278. The leading cause of fatal home accidents is
    A. fires
    B. explosions
    C. poisonings
    D. falls
    E. firearms

279. All of the following are minimum functions of a state health department, as defined by the American Public Health Association, **EXCEPT**

    A. the study of state health problems and planning for their solution

    B. coordination and technical supervision of local health activities

    C. financial aid to local health departments

    D. the enactment of sanitary regulations applicable in local health departments

    E. the establishment of maximum standards for local health work

280. The state health department is responsible for the protection of the general health and welfare of the citizens of that state. The supervision of these activities is usually carried out by the

    A. governor of the state

    B. state board of health

    C. commissioner or director of health

    D. state public health association

    E. directors of health planning agencies

Questions 281–294: Select the **INCORRECT** item or statement in each case.

281. The five leading causes of death in the United States in 1989 were, according to rank,

    A. heart disease

    B. cancer

    C. diabetes mellitus

    D. accidents

    E. chronic obstructive lung disease

282. Concerning cancer morbidity and mortality,

    A. a child born in the United States in 1985 had more than a 1 in 3 chance of eventually developing an invasive cancer

    B. for white women the probability of developing cancer of the breast during their lifetime is about 1 in 10

    C. currently it is projected that the chances of developing any form of cancer is greater for males of both races than for females

    D. the incidence of cancer of the prostate in males has apparently overtaken lung cancer

    E. in males of both races at age 65, the probability of lung cancer increases and prostate cancer decreases, especially in blacks

283. Concerning the sanitary quality of milk,

    A. the phosphatase test detects milk products that have not been adequately treated

    B. the phosphatase test is limited in that it cannot detect a batch of milk that has been contaminated with raw milk

    C. properly pasteurized milk generally has a bacteria count of less than 50,000/mL

    D. phosphatase and coliform tests collectively constitute a better index of safety of pasteurized products than the standard plate count

    E. the presence of coliform organisms in raw milk is of little significance

284. Concerning infant mortality in the United States,

    A. gastroenteritis is a rare cause of infant death today

    B. the infant mortality rate is much less than that found in Sweden and the Netherlands

    C. infant mortality rates show large differences between races

    D. high infant and maternal mortality rates occur among teenage mothers, especially the very young ones

    E. recent infant mortality rates (since 1990) run around 9.7/1000 live births, down from about 30/1000 in 1950

285. Regarding control and prevention of the spread of influenza by vaccine,
   A. toxicity (side effects) appears to be more frequently associated with immunization of older people (over age 70)
   B. inactivated vaccines are generally recommended for those with highest risk of fatal consequences
   C. formulations are changed frequently
   D. live virus vaccines are usually prepared with influenza virus A
   E. the efficacy of killed versus live influenza vaccine is a subject of continuing debate

286. Concerning methods of limiting family size,
   A. women who do not breastfeed are much more likely to become pregnant than those who do
   B. globally, the number of vasectomies far exceeds the number of tubectomies
   C. the barrier methods (condoms and diaphragms) are perhaps the most effective at limiting family size
   D. all anovulatory drugs are particularly likely to cause side effects in women over 35 years of age and women who smoke
   E. the theoretical effectiveness of IUDs (intrauterine devices) ranges from about 97% to 99%.

287. Legionnaire's disease
   A. occurs in both epidemic and sporadic form
   B. has an estimated incidence of 25,000 to 50,000 cases annually in the United States
   C. is rarely isolated from water sources unrelated to outbreaks of human disease
   D. is not controlled by routine testing of potable water systems or cooling towers for the causative agent, *Legionella pneumophilia*
   E. in epidemic form usually results from exposure of susceptible individuals to an aerosol generated by an environmental source of water contaminated with *Legionella*

288. Which of the following is advice for travelers to malaria epidemic areas?
   A. All travelers to malarious areas of the world are advised to take an appropriate drug regimen to prevent malaria
   B. Travelers can be advised that if an appropriate drug regimen is taken, they are safe from the likelihood of contracting malaria
   C. Persons who develop symptoms of malaria should seek prompt medical evaluation, including thick and thin malaria smears, as soon as possible
   D. Malaria symptoms can develop as early as eight days after initial exposure
   E. Malaria symptoms can appear months after departure from a malarious area, even after chemoprophylaxis is discontinued

289. Concerning mycobacterial infection in persons with human immunodeficiency virus (HIV) infection,
   A. the most common mycobacterial species isolated from patients with diagnosed AIDS is *Mycobacterium avium complex* (MAC)
   B. clinicians should consider the diagnosis of tuberculosis in patients with (or at risk of) HIV infection, even if the clinical presentation is unusual
   C. pulmonary tuberculosis in patients with HIV infection are easily distinguished from other pulmonary infections, such as *P. carinii* pneumonia
   D. chemotherapy should be started whenever acid-fast bacilli are found in a specimen from a patient with HIV infection and clinical evidence of mycobacterial disease
   E. individuals who are known to be HIV seropositive should be given a Mantoux skin test with five tuberculin units of PPD as part of their clinical evaluation

290. Concerning IUDs as a means of contraceptive,

    A. IUD users have about half the risk of unintended pregnancies as those relying on the condom

    B. wearing an IUD is absolutely contraindicated for women with active pelvic infection

    C. use of an IUD is contraindicated in those with uterine or cervical malignancy

    D. mortality rate for IUD users is about 3/million/yr

    E. uterine perforation is less likely to occur when an IUD is inserted into the uterus of a lactating woman

291. Concerning sterilization of both males and females,

    A. voluntary surgical sterilization is followed by about 0.4 unintended pregnancies per 100 procedures in the first year following surgery

    B. long-term adverse health effects associated with vasectomy continue to be of serious concern

    C. China and India exceed the United States (and probably the rest of the world) in the number of surgical sterilizations

    D. human studies have failed to confirm a serious risk of atherosclerosis in men who have undergone vasectomy

    E. globally, almost 100 million couples are reported to be using sterilization as a form of fertility control

292. Concerning death rates by race, sex, and age for the United States in 1985,

    A. white females had the lowest estimated age-adjusted death rate

    B. black females had the second lowest age-adjusted death rate

    C. black males had the highest age-adjusted death rate

    D. white males had age-adjusted death rates that were worse than black females, but better than black males

    E. there was little or no difference between the age-adjusted or crude death rates of any of these four groups.

293. With regard to mortality in relation to smoking,

    A. current male cigarette smokers have a 70% greater chance of dying from disease than nonsmokers

    B. specific mortality ratios are directly proportional to the amount smoked and the years of cigarette smoking

    C. mortality ratios are also higher for those who initiated their smoking at younger ages

    D. former cigarette smokers carry their increased mortality rate with them after they stop smoking

    E. after 15 years of no smoking, mortality rates for former smokers approach mortality rates of those who have never smoked

294. With regard to smoking and cancer of the lungs,

    A. the most definite causal relationship between the use of tobacco and any disease is with lung cancer

    B. lung cancer mortality rates in women are increasing much more slowly than in men

    C. cancer of the lungs has become the leading cause of cancer deaths in women in the 1990s

    D. the use of filter cigarettes with low tar and nicotine content decreases lung cancer mortality

    E. smokers experience decrease in lung cancer mortality rates, which approaches the rate of nonsmokers, after 10 or 15 years of cessation

DIRECTIONS (Questions 295–305): This section consists of clinical situations, each followed by a series of questions. Study each situation and select the **one** best answer to each question following it.

Questions 295–296: An illness characterized by fever, lymphadenopathy, and headache occurred among patrons of a riding stable. Twenty-nine people were ill. Most cases occurred in mid-October. Most of the 29 had serologic evidence of acute toxoplasmosis and an additional five persons at the stable had serologic evidence thereof, but remained asymptomatic. Of the 34 persons, 30 were identified as women, and 24 were between 16 and 30 years of age. Twenty people at another stable were interviewed and were tested for toxoplasmosis antibodies and all were negative. Cats from the stable associated with the outbreak were bled and serologic tests revealed that two of the three cats had elevated toxoplasma titers.

295. Based on this information, the most logical deduction is that
   A. women are much more likely to acquire toxoplasmosis than men
   B. symptoms of toxoplasmosis are diagnostic
   C. toxoplasmosis is mostly transmitted from horses
   D. cats are nontransmitters of toxoplasmosis
   E. the patrons acquired the illness from the cats

296. Other pertinent facts about toxoplasmosis include all the following **EXCEPT**
   A. pregnant women may transmit to their developing fetuses
   B. it may be transmitted by eating poorly cooked, or raw, infected meat
   C. studies show that 30% to 40% of the residents of the United States have serologic evidence of previous infection
   D. cats (and all Felidae) are the only animals excreting oocysts in their feces
   E. toxoplasmosis is also a venereal disease

Questions 297–299: A 35-year-old man returned to the United States from Thailand. He had a three-day history of chills and fever, headache and myalgia, weakness, and loss of appetite. He had lived in Thailand for the past 12 years, where he had had three episodes of malaria. He gave a history of exploratory laparotomy and splenectomy following an auto accident in the 1950s.

297. Based on the above information, the most important diagnosis to consider is
   A. malaria
   B. schistosomiasis
   C. influenzal syndrome
   D. trypanosomiasis
   E. smallpox

298. Ten days after the onset of the symptoms he was admitted to a hospital where a diagnosis of malaria was established. The only abnormal physical findings recorded were a 30-cm midline surgical scar on the abdomen, and left upper quadrant fullness believed to be a prominent left lobe of the liver. It is most likely that

    A. the surgical scar represents previous splenectomy

    B. left upper quadrant fullness represents gas in the splenic flexure

    C. upper quadrant fullness is represented by a prominent left lobe of the liver

    D. the patient has carcinoma of the transverse colon

    E. the clinical findings are of no significance to the present problem

299. A peripheral blood smear showed trophozoites and gameto-cytes of *Plasmodium vivax*. He was placed on a course of oral chloroquine phosphate. He rapidly became afebrile and asymptomatic. Shortly after admission, the patient fell and, following cardiopulmonary arrest, he died. An autopsy was done. It most likely showed

    A. a ruptured liver

    B. a ruptured diverticular

    C. a ruptured spleen

    D. a ruptured stomach

    E. findings that had no significance to the case

Questions 300–301: A 7-year-old female was admitted to a hospital with a diagnosis of acute cerebellar ataxia. Her symptoms were unsteady gait and inability to walk. Her temperature was 99 °F. A lumbar puncture revealed no WBCs; spinal fluid protein was 22 mg/mL and bacterial culture revealed no growth.

300. Based on this information, the most likely etiology is

    A. an infection

    B. an intoxication

    C. a stroke

    D. neoplasm

    E. psychosomatic complaint

301. Dog tick paralysis was suspected and after a search, two ticks (Dermacentor variabilis) were discovered attached to her scalp. There was obvious clinical improvement within 6 to 8 hours of removal of the ticks. Based on your knowledge of tick paralysis, what do you now suppose is the etiology of the illness?
- **A.** Bacterial infection
- **B.** Slow virus
- **C.** Toxin
- **D.** Rickettsia
- **E.** Protozoa

Questions 302–305: A group of 215 children is found to have hemoglobin values ranging from 9.0 to 14.9 g/100 mL, with a mean of 11.94 g/100 mL. The distribution is relatively normal with a standard deviation of 1.0 g/100 mL. On the basis of this study, one can conclude the following.

302. The average hemoglobin value for the group is
- **A.** 9.94g/100 mL
- **B.** 10.94g/100 mL
- **C.** 11.94/100 mL
- **D.** 12.94/100 mL
- **E.** 13.94/100 mL

303. The percentage of hemoglobin values larger than the average for the group is
- **A.** 20%
- **B.** 30%
- **C.** 40%
- **D.** 50%
- **E.** 60%

304. The percentage of hemoglobin values lying between plus one standard deviation and minus one standard deviation is
- **A.** 15%
- **B.** 25%
- **C.** 33-1/3%
- **D.** 68%
- **E.** 75%

305. The percentage of hemoglobin values less than minus 2 standard deviations from the mean is
   A. 1%
   B. 2%
   C. 4%
   D. 8%
   E. 16%

DIRECTIONS (Questions 306–330): Each group of questions below consists of lettered headings followed by a list of numbered words or statements. For each numbered word or statement, select the **one** lettered heading that is most closely associated with it. Each lettered heading may be selected once, more than once, or not at all.

Questions 306–310:
   A. Incidence rate
   B. Point prevalence rate
   C. Period prevalence rate
   D. Proportional mortality rate
   E. Neonatal mortality rate

306. Death under 28 days of age, divided by live births

307. Deaths assigned to a specific disease, divided by all deaths

308. New cases of a disease diagnosed during a given period, divided by mid-period population

309. Existing cases of disease at a single moment in time, divided by population at the same moment

310. Current cases, new and old, of a disease occurring during a given interval of the month, divided by the mid-interval population

Questions 311–315:
  A. Rabies
  B. Tetanus
  C. Rat-bite fever
  D. Leptospirosis
  E. Lymphocytic choriomeningitis

311. Worldwide it has an incubation period of 3 to 21 days; case fatality rate varies from 30% to 90%. Laboratory tests are of little help in diagnosis. Reservoir: intestine of man and animals

312. Incubation period in man is as short as ten days and up to one year; rapidly fatal paralytic infection; incidence uniform throughout the year. Rare airborne spread. All warm-blooded mammals are susceptible. Natural immunity in man is not known

313. Sudden onset with chills, fever, prostration, and myalgia; evidence of kidney involvement often present; fatality rate up to 20% in cases with jaundice and renal damage

314. Known also as Haverhill fever; infectious agent is *Streptobacillus moniliformis*; some local epidemics spread by contaminated milk or water

315. An endemic viral infection of animals transmissible to man; deaths are rare; organism is transferred from house mouse to man in feces and urine; nature of illness established by virus isolation

Questions 316–320:
  A. Motor vehicle fatalities
  B. Motorcyclists
  C. Persons under 20 years of age
  D. Fatalities on American highways
  E. Alcohol use

316. Sevenfold greater chance of fatal injury per mile than automobile riders

317. Decreased by 4% in 1981

318. Leading cause of lost years of potential life

319. Account for 30% of fatal motorcycle accidents

320. Accounted for 54,200 deaths in 1980

Questions 321–323:
  A. Fertility
  B. Fecundity
  C. Crude birth rate (CBR)
  D. Total fertility rate
  E. General fertility rate

321. The ability to produce live-born young

322. Number of live births/1000 population during a year

323. Provides a single number to summarize the fertility level

Figure 3.6

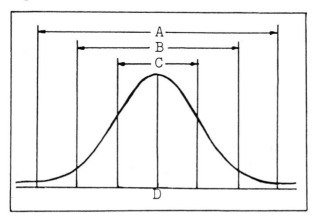

Questions 324–328: Refer to Figure 3.6.
- **A.** A
- **B.** B
- **C.** C
- **D.** D

324. The mean

325. $x - s$ to $x + s$

326. Laboratory standards are based on this value

327. Contains 95% of the data

328. Within three standard deviations of the mean

Questions 329–330: Refer to Figure 3.7.
- **A.** A
- **B.** B
- **C.** C
- **D.** D
- **E.** E

329. The mean is greater than the median

330. The distribution is skewed to the left

Figure 3.7

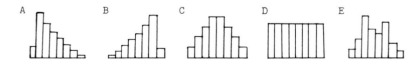

DIRECTIONS (Questions 331–350): For each of the questions or incomplete statements below, **one** or **more** of the answers or completions given is correct. Select

A   if only **1, 2,** and **3** are correct,
B   if only **1** and **3** are correct,
C   if only **2** and **4** are correct,
D   if only **4** is correct,
E   if **all** are correct.

---

331. In the absence of a planned experiment, a judgment about the causal nature of an association is made on which of the following?
1. Strength of the statistical association
2. Time sequence
3. Consistency with existing knowledge
4. Clinical impression

332. The National Health Survey was initiated in 1957 to obtain accurate and systematic information in the United States of illnesses in the population as a whole. Which of the following is(are) applicable to the preceding statement?
1. It is the responsibility of the National Center for Health Statistics (NCHS)
2. The health interview survey gets its information from the Framingham study
3. The health examination survey component is based on national population samples and the results are published in individual monographs
4. The survey focuses almost exclusively on geriatric health problems

333. Sometimes it is necessary to use a rate to compare health and disease between two populations (communities). Which of the following is(are) applicable to the preceding statement?
1. Such comparisons can be misleading without standardization
2. There is a direct method of standardization
3. There is an indirect method of standardization
4. It is seldom necessary to standardize for age

334. Which of the following is(are) true concerning a null hypothesis (H_0)?
    1. Any differences found in the statistics being compared are the result of random sampling
    2. Test of a null hypothesis is subject to two types of errors
    3. Type 1 error is a result of rejecting the null hypothesis when it is in fact true
    4. Type 2 error is a result of failing to reject the null hypothesis when it is in fact false

335. Which of the following is(are) correct concerning the mode of transmission of meningococcal meningitis?
    1. No extrahuman reservoirs
    2. Person-to-person transfer with portal of exit through the nasopharynx
    3. Probably respiratory droplets, although air-borne under certain conditions
    4. Association of disease with crowded living conditions

336. Influenza is characterized epidemiologically by
    1. epidemic-pandemic potentiality
    2. mortality among predictably high-risk groups
    3. periodic-cyclic recurrences
    4. high morbidity–low mortality

337. Diseases with known genetic basis that can be ameliorated or corrected by therapeutic or surgical means include
    1. spina bifida
    2. sickle cell anemia
    3. Turner's syndrome
    4. phenylketonuria (PKU)

338. Examples of federally sponsored programs for health include
    1. Hill-Burton program
    2. regional medical programs (RMP)
    3. comprehensive health planning (CHP)
    4. health systems agency (HSA)

Directions Summarized				
A	**B**	**C**	**D**	**E**
1,2,3	*1,3*	*2,4*	*4*	*All* are
only	only	only	only	correct

339. Which of the following is(are) correct with reference to Figure 3.8?
1. This data indicates a secular change in the number of sterilizations
2. This is period prevalence data
3. If a secular change is present, it could relate to over- or under-reporting, which may be good or poor
4. The data is given in rates

Figure 3.8 Surgical sterilization. Tubal sterilizations performed in U.S. hospitals, 1970–1978. Reproduced from Morbidity and Mortality Weekly Report, Annual Summary 1980, Vol. 29, No. 54, September 1981, U.S. Department of Health and Human Services, Public Health Service.

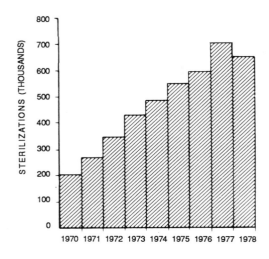

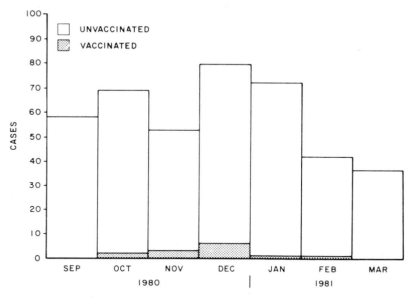

Figure 3.9 Guillain-Barré syndrome. Reported U.S. cases by month of onset of neurologic symptoms, 1980–1981. Reproduced from Morbidity and Mortality Weekly Report, Annual Summary of 1980, Vol. 29, No. 54, September 1981, U.S. Department of Health and Human Services, Public Health Service.

340. In 1978, CDC established a surveillance system to detect cases of Guillain-Barre syndrome (GBS). The system was to determine whether or not an increased risk of influenza vaccine-associated GBS existed for the influenza vaccines administered in 1978 through 1979. Figure 3.9 shows the total number of vaccinated and unvaccinated cases for 1980–1981. Which of the following is(are) pertinent to this information?
1. The diagram indicates that there was no association between the administration of the vaccine and the development of GBS
2. There is good evidence in this data of a noncausal association
3. This is period prevalence data
4. It is not possible to tell from this data whether there is a causal association between the vaccine and the development of GBS

Directions Summarized

A	B	C	D	E
1,2,3	*1,3*	*2,4*	*4*	*All* are
only	only	only	only	correct

341. In the United States, Rocky Mountain spotted fever is passed to man by the bite of appropriate ticks. Study Figure 3.10 concerning the distribution of RMS in the United States in 1980. Which of the following is(are) correct?
    1. The data is given in rates and consequently is appropriate for comparisons with other times and other places
    2. This is a spot-map; the data is given in raw numbers, or groupings thereof
    3. This data is relatively unaffected by physicians' reporting habits
    4. Despite the name of the disease, a physician is more likely to encounter Rocky Mountain spotted fever in the eastern part of the United States than in the West

342. During a measles epidemic, it is advisable to
    1. isolate patients for 14 days after the rash appears
    2. close schools
    3. give gamma globulin in prophylactic doses to all children who are direct contacts
    4. recognize that quarantine is impractical

343. Early syphilis differs from late syphilis in which of the following respects?
    1. Infectiousness
    2. Destructiveness
    3. Reinfection rate
    4. Serologic reaction

Figure 3.10 Typhus fever, tickborne (Rocky Mountain spotted fever). Reported cases by U.S. county, 1980. Reproduced from Morbidity and Mortality Weekly Report, Annual Summary 1980, Vol. 29, No. 54, September 1981, U.S. Department of Health and Human Services, Public Health Service.

344. Which of the following is(are) correct concerning vaccination during pregnancy?

1. Pregnancy is a contraindication for the use of vaccines against rubella, measles, and mumps

2. If a vaccine must be given during pregnancy, it is wise to wait until the second or third trimester, whenever possible

3. Vaccinating pregnant women with killed (inactivated) vaccine does not augment the risk to the fetus

4. Tetanus and diphtheria toxoid should be given to inadequately immunized pregnant women because this affords protection against neonatal tetanus

Directions Summarized				
A	**B**	**C**	**D**	**E**
1,2,3	*1,3*	*2,4*	*4*	*All* are
only	only	only	only	correct

345. Cases of AIDS continue to proliferate and
1. currently reported AIDS cases have resulted from HIV exposure up to seven years before diagnosis
2. the possibility of longer incubation periods cannot be excluded
3. due to the long period between infection and the development of AIDs, transfusion-associated cases are expected to continue
4. persons meeting the AIDS case definition are only a small percentage of all persons infected with HIV

346. Which of the following is false concerning hepatitis B and the delta agent?
1. Combined hepatitis B virus and delta virus infection is recognized to cause hepatitis outbreaks with unusually high mortality in parenteral drug abusers (PDSs)
2. Delta superinfection of hepatitis B virus carriers frequently causes transformation from no or mild chronic liver disease to severe progressive chronic active hepatitis
3. In the United States, infection with the delta agent has been observed previously among persons with frequent blood (and blood products) exposures
4. In the United States, almost 100% of persons reported with hepatitis B infection have been identified as parenteral drug abusers

347. Children with symptomatic HIV infection have immunologic abnormalities similar to those of adult AIDS patients, including which of the following?
1. Hypergammaglobulinemia
2. Decreased T4 lymphocytes
3. Reversed helper/suppressor T-cell ratios
4. Leukopenia

348. Which of the following occupational agents is(are) associated with lung cancer?
1. Arsenic
2. Asbestos
3. Nickel
4. Ionizing radiation

349. Which of the following statements concerning toxic fumes and dusts is(are) true?
1. Silo-fillers disease is a toxic pulmonary edema which follows inhalation of the oxides of nitrogen in freshly filled silos
2. Organic dust toxic syndrome (ODTS) is a generic designation to describe a syndrome different from silo-fillers disease and silo-unloaders syndrome
3. "Mill fever" is an ODTS-like pulmonary condition found in cotton textile workers
4. "Humidifier fever" is an ODTS-like syndrome found in building occupants exposed to air from highly contaminated ventilation systems

350. An important concept in the Medicare program is the benefits period. Which of the following is(are) applicable?
1. The first benefits period starts as soon as one turns 65
2. The benefits period can last for no longer than 60 days
3. One can have only six benefits periods in his lifetime
4. A benefits period is a way of measuring one's use of services under Medicare hospital insurance

Explanatory Answers

235. E. Life tables are not directly associated with the chi-square test of statistical fame, but qualitative data are. Degrees of freedom must be employed correctly. The null hypothesis and a significance level are both "part and parcel" of a chi-square test. (Ref. 1, p. 142)

236. A. Associations necessary for appropriate use of a z test include a random sample with a normal distribution, quantitative rather than qualitative data, and a known population standard deviation. The correlation coefficient (r) is a biostatistical function of itself and is not dependent on the z test. (Ref. 1, pp. 81–83)

237. E. In performing a significance test the null hypothesis must be stated and eventually be accepted or rejected. A p value must be calculated and conclusions must be drawn on the basis of the value of p. Standardization (or adjustment) of rates is a procedure to permit valid comparisons between rates found in two distinct populations, one of which is to be compared with the other. (Ref. 1, p. 106)

238. B. Prospective studies in the main are costly. They follow large numbers of individuals over long periods of time. Advantages are plenty, however, for they do lack bias in factor, they do yield incidence rates as well as relative risk. Also, they may yield associations with additional disease as a by-product and they are efficient for studying rare exposure. (Ref. 2, p. 169)

239. D. In fact, a disadvantage to retrospective studies is that they cannot yield rates, only ratios. (That is, relative risk.) Other disadvantages are the bias of recall and the problem of selecting control groups. It is true, however, that they are relatively inexpensive, relatively rapid, and consequently can be easily repeated. (Ref. 2, pp. 163, 169)

240. A. The spores are ubiquitous in soil and can contaminate fresh foods, particularly those harvested from the ground. Spores are heat resistant and can survive boiling for hours. To germinate, several conditions must be met, including the appropriate pH and oxygen content. Food-borne botulism generally results from home-canned vegetables that are contaminated with spores and are improperly prepared. The toxin is heat labile. (Ref. 35)

241. D. Bacterial contamination is combatted by the addition of a disinfectant, chlorine (not ·fluorine) in a concentration of 0.2 to 0.5 parts per million. Fluorine may be added to prevent dental caries, and the concentration is usually in the range of 0.7 to 1.5 parts per million. Primary sedimentation and clarification is accomplished by storage reservoirs and exposure to air and light. A coagulant, usually alum, is used to remove residual turbidity and then passage through a rapid filter is appropriate. Contamination of purified water can be prevented by residual disinfection of the distribution system and the prevention of cross connections and back syphonage. (Ref. 9, p. 366)

242. B. Salvage and composting together only account for about 6% of waste disposal. Open dumping accounts for 73%. Consequently, most urban communities are running out of places to dump or bury. Cost of disposal is considerable and about two thirds of it is represented by pick-up and transfer to a nearby collection vehicle. The 3.5 billion tons include paper, garbage, glass, ceramics, metals, wood, cardboard, textiles, plastic, rubber, and leather. (Ref. 9, p. 370)

243. E. The process includes the chemical addition of the aluminum sulfate, rapid mixing, and coagulation. This is to remove the finely divided suspended material (called colloid material), microorganisms, and, to some extent, dissolved substances, by bringing them together into flocks sufficiently large enough to be removed by sedimentation and/or filtration. (Ref. 7, p. 822)

244. B. The main purpose of sludge treatment is dewatering and destruction of organic matter. This is usually done by digestion, settling, trickling, etc. (Ref. 7, p. 840)

245. E. In the United States and other industrialized countries, obesity is probably the most prevalent form of malnutrition. The true medical importance of obesity has only recently become appreciated. (Ref. 7, p. 1520)

246. A. Oysters are more frequently eaten raw and represent the largest harvest of the shellfish. They are more often involved in disease outbreaks. Man has made oysters, clams, and other shellfish a danger to himself by polluting the waters in which they live. Typhoid fever and other intestinal infections, including viral hepatitis, have been convincingly traced to infected oysters and clams. (Ref. 7, p. 180)

247. B. Tetanus spores are found in the GI tracts of many animals, humans not excluded. The agent is spore-forming and is anaerobic. Early cleansing of wounds with removal of necrotic and nonviable tissue has, by military experience, been demonstrated as valuable. This is not so with antibiotics and disinfectants. Tetanus toxoid, however, is one of the most effective of all immunologic agents. (Ref. 7, pp. 432–435)

248. D. Herpes simplex is caused by a herpes virus. All of the others are adenovirus associated. (Ref. 7, pp. 152–153)

249. A. These cases are incidence data. They do not represent duration, so they cannot be either point or period prevalence. It is believed that water-borne diseases are grossly under-reported, not over-reported. (Ref. 5)

250. A. The authors are correct, for the fallacy of comparing percentages is encountered only if one is comparing a series of percentages with another series of percentages. This is simply a single series. "Fallacy of the average" is not a consideration in this data. Controls are not necessary and this is not a normal distribution. (Ref. 6, p. 2)

251. B. In managing dirty wounds, TIG is only recommended for those who have had unknown or <3 doses of tetanus toxoid. Those who have not had a dose of tetanus toxoid within ten years also require TIG. (Ref. 14)

252. E. Each of the following vaccines should be given intramuscularly: DTP, hepatitis B, influenza, plague, tetanus, TD, DT, and *H. influenzae* b conjugate vaccine. HbPRP (*H. influenzae* b polyribosylribitol-phosphate vaccine) and pneumococcal vaccines may be given subcutaneously or intramuscularly. (Ref. 15, p. 207)

253. C. The Immunization Practices Advisory Committee recommends that there be a minimum of a four-week interval between the administration of two live antigens if they are not administered simultaneously. Two or more killed antigens or a combination of a killed and a live antigen may be given simultaneously or at any interval between doses. (Ref. 15, p. 219)

254. E. Although a moderate to severe febrile illness is reason to postpone immunization, a minor illness such as a mild URI with or without low-grade fever is not. A fever of above 105 °F should not be considered minor and would warrant postponement of vaccination, and probably investigation and treatment. (Ref. 15, p. 223)

255. D. Significantly immunocompromising conditions, including congenital and acquired immunodeficiency with HIV infection, leukemia, lymphoma, generalized malignancy, certain severe autoimmune disorders, cancer chemotherapy and radiation therapy, and high-dose corticosteroids should be given the following vaccines: influenza, pneumococcal, and hepatitis B X 2. Live vaccines should not be given. Exceptions do exist for certain HIV and leukemia patients. (Ref. 16)

256. B. The recommended schedule for active immunization of normal children and infants is shown in Table 3.2. (Ref. 15, p. 210; Ref. 18)

Table 3.2

Recommended Age	Vaccine	Comments
2 mo	DTP, OPV	OPV and DTP can be given earlier in areas of high endemicity
4 mo	DTP, OPV	6-wk to 2-mo interval desired between OPV
6 mo	DTP	An additional dose of OPV is optional in areas of high risk of polio virus exposure
15 mo	DTP, OPV, MMR	MMR should not be given to children less than 12 mo
18 mo	HbCV	Conjugated preferred over polysaccharide
4–6 yr	DTP, OPV, MMR,	At or before school entry
14–16 yr	TD	Repeat every 10 years throughout life

257. B. See Table 3.3. (Ref. 17, p. 317)

Table 3.3

Abbreviation	Term	Comments
HBV	Hepatitis B virus	Surface antigen of HBV
HBsAG	HB surface antigen	Surface antigen of HBV detectable in large quantity of serum; correlates with HBV
Anti-HBs	Antibody to HBsAG	Indicates past infection with and immunity to HBV, passive antibody from HBIG vaccine, or immune response from HBV vaccine
IGM Anti-HBc	IGM class antibody to HBcAG	Indicates recent infection with HBV; positive for 4–6 mo after infection

258. D. Because of recent outbreaks of measles and fluctuating immunity, statements in B, C, and E are being emphasized. Furthermore, 2 doses of vaccine are recommended instead of one. (Ref. 18)

259. A. The most important values in describing the frequency distribution of a series of observations are the mean and standard deviation. (Ref. 1, pp. 36–40)

260. E. When considering a normal distribution, 99.73% of your observations will be within three standard deviations (SD) from the mean. Values more than this are very rare. (Ref. 1, p. 68)

261. A. Plus or minus three SD contain 99.73%; plus or minus two SD contains 95.45%; plus or minus one SD contains 68.27%. (Ref. 1, p. 68.)

262. C. By plotting the data, one can determine the strength of the linear correlation between y and x. A positive r, as seen in the example, indicates that as y increases, x will increase. (Ref. 1, pp. 162–163)

263. A. The proportion of observations that lie within plus or minus one SD from the mean is 68%; two SD is approximately 95.5%; and three SD is 99.73%. (Ref. 1, p. 68)

264. A. A positive linear correlation is seen when y increases and x increases. A negative correlation is when y decreases and x increases. (Ref. 1, p. 165)

265. E. The ideal normal frequency distribution characteristics are: (1) the mean, median, and mode all coincide; (2) the curve is perfectly symmetrical around the mean; and (3) the standard deviation will contain a known percentage of observations. Range is the difference between the largest and smallest pieces of data. (Ref. 1, p. 39)

266. C. The difficulties in controlling the snail population make it hard to eliminate schistosomiasis. (Ref. 7, pp. 453–454)

267. C. The cholera vibrio proliferates within the lumen of the intestines and does not invade the bloodstream or the tissues. A powerful endotoxin is elaborated by the organism, which produces an inhibition of sodium transport by the cells of the intestinal mucosa, thus impairing its absorption capacity. The result is loss of electrolytes. The fluid lost is mainly isotonic. (Ref. 7, p. 244)

268. A. Rhinovirus, of which there are many recognized serotypes, plus at least one subtype, are the major known etiologic agents of the common cold in adults. (Ref. 4, p. 327)

269. E. When talking about epidemics and pandemics, influenza is at the forefront. It is known to cause mortality among predicted high-risk groups (infants, young children, and those over 65). An epidemic causes excessive morbidity, but low mortality. Because of the periodic-cyclical recurrence, epidemiologists can forecast (with limited accuracy) future epidemic patterns and predict the likely prevalent virus type. Winter prevalence is most likely. (Ref. 7, pp. 142–144)

270. B. Oral contraception, the pill, is the most effective means of temporary contraception. National survey data show that the pill has a failure rate of 2.4 unintended pregnancies per 100 married sexually active women during the first year of use. The risk of pregnancy for women who use oral contraception is half that of women who wear IUD's. An estimated 10,000,000 women in the United States used oral contraceptives in 1982. It is the most popular method of birth control for women younger than 30 years of age. (Ref. 7, p. 1595)

271. C. Estrogen containing contraceptive pills are not prescribed for women during the first six weeks that they are breastfeeding an infant. Other reasons for contraindicating the pill are: unexplained vaginal bleeding, pregnancy, stroke, hypertension, gallbladder disease, diabetes, plans for surgery within four weeks of taking the pill, and injury or embolization of the legs. The risk of problems increases with cigarette smoking, age, and the use of pills that have a high estrogen content. (Ref. 7, p. 1597)

272. B. The tendency for the number of births in a population to continue increasing even after a fertility decline has set in is referred to as population momentum. If a population has been growing at 2.5% per year with a constant crude birth rate for a long time, its number of births was increasing at 2.5% per year. Thus, if mortality rates stay relatively constant, the number of persons reaching age 15 will continue to increase at 2.5% for 15 years into the future, no matter what happens to fertility rates in the meantime. The implications for developing countries is obvious, for no matter how rapidly fertility rates decline to replacement levels, these countries still have many years of rapid population growth ahead. (Ref. 7, p. 83)

273. D. Increased abortion among first premarital pregnancies is true. All of the others are false. There is, however, an increase in the proportion of unmarried teens keeping their child once it is born. These facts raise questions about the adequacy of current sex education and family planning services for teenagers, and about changing social patterns among teens (Ref. 7, p. 96)

274. E. Fluoridation of municipal water supplies is a common practice to prevent dental caries. A fluoride ion level of 0.6 to 1.2 mg/L (depending on temperature) will prevent dental caries with no mottling of enamel. (Ref. 7, p. 1480)

275. A. Regarding all causes of death, toxemias lead the list for maternal mortality. (Ref. 29, p. 463) [And maternal mortality runs less than 400 deaths per year in the United States—Ed.]

276. A. Epilepsy appears to involve complex interactions (as yet unanalyzed) between genetic and nongenetic causal factors. (Ref. 30, p. 323)

277. B. Low back pain is the most expensive occupational health problem in the United States. (Ref. 3, p. 342)

278. D. Falls account for almost half of the accidental deaths in the home. Fire burns rank second. (Ref. 7, p. 1558)

279. E. The state health department is required to establish minimal, not maximal, standards for local health work. The other functions are: the maintenance of central and branch laboratory services, including diagnostic, sanitary, chemical, biological, and research activities; the collection, tabulation, and analysis of vital statistics; the collection and distribution of information concerning preventable disease; the maintenance of a safe quality of water and the control of waste disposal; establishment and maintenance of minimal standards of milk sanitation; provision of service to aid industry in the control of occupational hazards; the establishment of qualifications for health personnel and formulation of plans in cooperation with other organizations for meeting all health needs. (Ref. 37)

280. C. The commissioner or director of health is the administrative head of the state health activities. His line of authority is usually from the governor, with the state board of health service serving mainly as an advisory board. (Ref. 37)

281. C. Diabetes mellitus is the seventh in rank. The correct order is heart diseases, cancer, cerebrovascular diseases, accidents, and chronic obstructive lung disease. Number 6 is pneumonia and influenza. Heart diseases account for 36.4% of all deaths, cancer 22.3%, and cerebrovascular diseases 7.1%. Thus, the top three account for about two-thirds (65.8%) of all deaths. (Ref. 10, p. 12)

282. E. By the age of 65, the probability of lung cancer decreases, but prostate cancer continues to increase, affecting older men, and especially blacks. Each of the other statements is correct. The reader should be cautious to discern between cancer morbidity and cancer mortality when interpreting these statements. (Ref. 10, pp. 11–15)

283. B. The addition of raw milk to that which has already been treated and tested can be detected by the phosphatase test and that is one of the reasons for it. The presence of coliform organisms in raw milk is of little significance because contamination with these organisms is practically unavoidable. Coliforms in pasteurized milk would indicate contamination (postpasteurization contamination) or inadequate pasteurization. (Ref. 7, pp. 771–772)

284. B. In comparison with Sweden and the Netherlands, the United States' infant mortality rates are much higher. In 1990 the United States' rate was 9.7 per 1000 live births, Sweden's 5.8, and the Netherlands 7.6, respectively. The difference in IMR between races is related to a large extent to the differences in incidence of low birth weight. (Ref. 39)

285. A. Toxicity seems to be more frequently associated with younger persons (ie, those below the age of 20). During use of a swine influenza virus vaccine (1976), the Guillain-Barre syndrome was reported to affect one in each 120,000 vaccinees. It is not yet known whether this syndrome is associated with other influenza vaccines. (Ref. 7, pp. 144–145)

286. C. Barrier methods are less effective than other methods because they are inconvenient and because they require thought and action at the time of intercourse. All of the other statements are correct. (Ref. 7, pp. 1595–1601)

287. C. *Legionella* is frequently isolated from water sources unrelated to outbreaks of human disease. Thus, routine testing of potable water systems or cooling towers is not of much value and, therefore, is not recommended, since the organism is ubiquitous. In previous outbreaks, disease has been associated with exposure to evaporative condensers, cooling towers, showers, whirlpools, and even respiratory therapy equipment. (Ref. 19, p. 349)

288. B. Travelers must be informed that regardless of the antimalarial regimen employed, it is still possible to contract malaria. Common symptoms of malaria are headache, malaise, fever, chills, and sweats, which may occur at intervals. Neither physician nor traveler should presumptively ascribe the symptoms of a "flu-like" illness. Delaying appropriate treatment can have serious or even fatal consequences. (Ref. 36, p. 3)

289. C. Pulmonary tuberculosis in patients with HIV infection cannot readily be distinguished from other pulmonary infections. The other statements are true. With some exceptions, patients with tuberculosis and HIV infection respond relatively well to standard antituberculosis drugs. Their treatment, however, should include at least three drugs initially, and the treatment may need to be longer than the standard duration of nine months. Some false negative skin test results may be encountered in providing HIV infection seropositive patients. They should be given a Mantoux skin test anyway, for significant results are still meaningful. (Ref. 13)

290. E. It is more likely to occur if insertion takes place during lactation. Three conditions—pelvic inflammatory disease, uterine perforation, and second trimester septic spontaneous abortion—account for virtually all of the hospitalizations in users of IUDs. (Ref. 7, p. 1597)

291. B. Long-term adverse health effects associated with vasectomy are not of the serious concern they once were because human studies have failed to confirm animal studies. Sterilized monkeys who were fed high cholesterol diets showed more serious atherosclerosis than those monkeys that had not been sterilized. China reported about 36 million sterilizations in 1980, and India about 23 million. (Ref. 7, p. 1601)

292. E. The differences were as follows:

White females: 390 deaths/100,000 population
Black females: 585 deaths/100,000 population
White males: 690 deaths/100,000 population
Black males: 1029 deaths/100,000 population

(Ref. 28, p. 6) [Figures are rounded for simplicity.—Ed.]

293. D. Former cigarette smokers experience declining mortality rates as the years of not smoking increase. Overall mortality rates for female smokers are less than for male smokers. However, tests of females with smoking characteristics similar to those of men experience mortality rates similar to those of male smokers. Mortality rates are decreased in smokers who use cigarettes with decreased tar and nicotine content. (Ref. 38, pp. 38–41, 43–46)

294. B. Lung cancer mortality rates are increasing more rapidly in women than in men, and this is why lung cancer has now surpassed cancer of the breast as the leading cause of cancer deaths in women. (Ref. 38, p. 21)

295. E. Both sexes are susceptible to toxoplasmosis. The symptoms are not particularly characteristic, generally limited to fever, lymphadenopathy, and headache. It is believed that toxoplasmosis is transmitted from infected cat feces, rather than from horses. (Ref. 32)

296. E. Transmission to fetuses is possible and may cause severe neurological and ocular abnormalities. An outbreak occurred among five medical students who ate raw hamburger. Studies do show that up to 40% of U.S. residents have serologic evidence of previous infection. Cats do excrete oocysts in their feces and, interestingly enough, only at one time in their lives and only for a period of approximately two weeks. We have found no cases to suggest that toxoplasmosis can be passed from person to person in the venereal mode. (Ref. 32)

297. A. While many diagnoses could give symptoms similar to those described, malaria must be first on one's list of clinical suspicions for no other reason than the good potential for cure from adequate treatment. (Ref. 33)

298. A. In view of the history, the surgical scar can certainly represent the area of exploratory laparotomy and the removal of the spleen. Carcinoma of the transverse colon could cause the mass, but is uncommonly the cause of the other symptom complex. The left upper quadrant fullness would typically support an enlarged spleen, a symptom not unusual with a diagnosis of malaria. (Ref. 33)

299. C. The spleen had ruptured and the peritoneal cavity contained 4 L of blood. Splenectomy commonly occurs in malaria; but rupture of the spleen occurs only rarely and usually follows an episode of acute *P. vivax*. The patient had reported the removal of his spleen. Was he incorrect? Not necessarily so. He may have had an accessory spleen removed; the latter are not rare and in one study they were found to be in 10% of 3000 routine autopsies. (Ref. 33)

300. B. An infection is unlikely because of the absence of WBCs and the negative bacterial culture. Viral infection, however, cannot be ruled out on the information given. Stroke is unlikely at the child's age. We need more information to diagnose a neoplasm, particularly duration of symptoms. A psychosomatic complaint, while possible, should not be our first choice in a child who apparently was ill enough to admit to the hospital. (Ref. 34)

301. C. Tick paralysis is a toxin-induced disorder, associated with the attachment and feeding of the female member of the tick species. (Ref. 34)

302. C. The "average" in statistical terms is known as the "mean." (Ref. 1, p. 38)

303. D. With this large sample size, 50% will have values larger than the average. (Ref. 1, p. 71)

304. D. A standard deviation of one above and below the mean contains 68% of the total sample size. Plus or minus two standard deviations contains approximately 95.5%; and plus or minus three standard deviations contains approximately 99.7% of the total sample size. (Ref. 1, p. 71)

305. B. Approximately 2% of the sample is below a minus two standard deviation from the mean. (Ref. 1, p. 71)

306. E. Incidence rate is based on new events (ie, new cases of a
307. D. disease diagnosed during a given period). Point prevalence
308. A. rate is based on cases of a disease existing in a theoretical
309. B. moment of time. Period prevalence rate is based on current
310. C. cases, new and old, occurring in a given time interval. The proportional mortality ratio is defined as each death assigned to a specific disease, divided by all deaths. Finally, neonatal mortality rate is death under 28 days of age divided by live births. In each definition, the denominator is the population at risk. The only way to get into the numerator is by acquiring the condition. (Ref. 8, pp. 56–57)

311. B. Tetanus (lockjaw) is an acute disease brought about by toxin of the tetanus bacillus growing anaerobically at the site of injury. Common characteristics are painful muscular contractions, mainly in the masseter and neck muscles, and secondarily in the trunk muscles. Fatality rates vary from 30% to 90% (in the absence of effective immunization.) The best protection is by active immunization with tetanus toxoid. (Ref. 4, pp. 384–385)

312. A. Rabies usually develops into fatal acute encephalomyelitis. It progresses to paresis or paralysis. Incubation period is usually 2 to

8 weeks, depending on the extent of the laceration and location. (Ref. 4, pp. 310–312)

313. D. Leptospirosis is a group of diseases with protean manifestations. The usual characteristics are fever, headache, chills, severe malaise, vomiting, muscular aches, and conjunctivitis, infrequently renal insufficiency and jaundice. Fatality rates increase with age and reach 20% or more in patients with kidney involvement and jaundice. (Ref. 4, p. 214)

314. C. Rat-bite fever (that occurs in the United States is also called Haverhill fever), is caused by *S. moniliformis*. Common characteristics are abrupt onset with chills and fever, headache, and muscle pain, followed by maculopapular rash mainly on the extremities. Actual contact with rats is not needed. Transmission can be by secretions of infected animals. (Ref. 4, p. 319)

315. E. Mice are a common carrier for this virus. The characteristics are diversified with influenza-like attacks or meningeal symptoms. The virus is transmitted by excreted urine, saliva, and feces in food or dust eaten by man. (Ref. 4, pp. 222–223)

316. B. Motorcyclists. The protective effects of helmets is evident at all levels of injury severity and the degree of protection increases with severity. A nonhelmeted rider is twice as liable to acquire the minor head injury as a helmeted rider and approximately five times as likely to acquire a severe or critical injury. (Ref. 12)

317. D. The National Transportation Safety Board estimates that all road fatalities on American highways decreased by 4% in 1981, reversing a five-year trend. Each day, 145 persons, including 12 children, die in vehicular collisions. (Ref. 12)

318. A. Motor vehicle fatalities. In 1980 motor fatalities accounted for 54,200 deaths. (Ref. 12)

319. C. By 1975, as a result of federal requirements, all but three states had enacted laws requiring helmet use for motorcycle users. In 1976 the federal requirements were repealed, and by 1982 nine states had no helmet laws, and 22 had amended theirs to require helmets only for teenaged riders. (Ref. 12)

320. A. In 1980, approximately 90,000 children under six years of age and 800,000 children 6 to 16 years of age were injured by motor vehicles. (Ref. 12)

321. B. Fecundity is the ability to produce. (Fecundity is difficult to measure since it refers to the theoretical ability of a woman or couple to conceive and carry a fetus to term.) (Ref. 7, p. 75)

322. C. Crude birth rate can be shown as follows:

$$CBR = \frac{\text{Live births in 1 year}}{\text{Mid-year population}} \quad x \quad 1000 \text{ live-born young}$$

(Ref. 7, p. 76)

323. D. Total fertility rate provides a simple index of the fertility level. (Ref. 7, p. 77)

324. D. The mean (the average of the set of numbers) is always located in the middle of a normal distribution. (Ref. 1, p. 39)

325. C. Plus or minus one standard deviation from the mean would contain 68% of the data. (Ref. 1, p. 71)

326. B. Most "normal" laboratory tests are based on evaluation of a "normal" population. The standards are then based on 95% of this normal population. This would be within two standard deviations of the mean. (Ref. 1, p. 71)

327. B. The area containing 95% of the data is $x - 2s$ to $x + 2s$. (Ref. 1, p. 71)

328. A. Within three standard deviations ($x - 3s$ to $x + 3s$) contains 99.7% of the population. (Ref. 1, p. 71)

329. A. If you sampled a population of women and found that most of the women were within "normal" weight limits, but you had more women who were overweight than underweight, then most of your population would be centered around the mean but the median based on the number of women in the study would be somewhere to the right of the mean because of the larger portion of overweight women to underweight ones. (Ref. 1, p. 39)

330. B. A skewed graph is indicated when one tail is stretched out longer than the other. The longer tail is the identified side, therefore graphs A and B are skewed right and left, respectively. (Ref. 1, p. 39)

331. A. Strength of the statistical association implies a small P value. The more frequently the alleged cause and effect are observed together, by comparison with the expected frequency, the more likely is the association between them explained by a causal relationship. The cause should precede the effect. The existence of a series of known cause-and-effect associations will make an overall association seem not only understandable, but even probable. Clinical impression may be valuable to suggest early hypotheses, but it is usually inadequate of itself to provide confidence in a proposed cause–effect relationship. (Ref. 8, p. 18)

332. B. The National Health Survey is the responsibility of NCHS. There are two major components, the health interview survey and the health examination survey. The former is based on a national random sample of about 130,000 noninstitutionalized persons each year. The Bureau of the Census selects the sample and conducts the field interview. The responsibility for interview content, analysis, and publication rests with NCHS. The health examination survey was started in 1960 to fill the gaps in the health interview survey. It may focus its attention on any age group and the results are published in individual monographs. (Ref. 8, pp. 30–31)

333. A. Rates may vary between communities simply because of different distributions of the populations by age or sex (or other variable). Age differences between two populations being compared is a very common variable for which standardization (or adjustment) must be made. (Ref. 8, pp. 41–42)

334. E. All of the statements are correct. Statistics being compared are the result of random sampling coming from the same population;

thus, any difference between them is due to chance. If a null hypothesis is true, then a type 2 error is nonexistent and irrelevant. If the null hypothesis is false, type 1 error has no meaning. Invariably, we want studies that are designed to limit both types of errors. (Ref. 11, p. 4)

335. E. Meningococci found only in humans are spread from nasopharynx of one person to another. Transmission is most intense in closed, crowded conditions. Meningococcal meningitis is introduced into the household by an adult and spreads to older children and then to infants. Hospital contacts are infrequently colonized. (Ref. 7, p. 210)

336. E. Influenza is generally not a fatal disease in the United States. While it is rarely fatal, because it is so extensive, the total number of deaths can be large. (Ref. 7, pp. 142–143)

337. B. Sickle cell anemia can be treated with transfusions and phenylketonuria with special diet. (Ref. 7, p. 1431)

338. E. Hill-Burton was designed to promote hospital construction. RMPs were directed at the nonprofit, private sector, and centered in medical schools (with emphasis on heart, cancer, and stroke). CHP programs were to help integrate the many new health service support programs voted by Congress in the mid-1960s. HSA was specifically directed to population-based health planning in local communities. (Ref. 7, pp. 1681–1682)

339. B. This does represent a secular change, an increase over the period 1970 to 1978. It is incidence data, for each tubal sterilization represents a new event. The secular change could be due to over-reporting or under-reporting, and it would be necessary for us to know the source of the data and the soundness (or completeness) of its gathering in order to determine its veracity. (Ref. 31, p. 122)

340. D. While the two variables (influenza vaccine and Guillain-Barré syndrome) may be associated in time and place, this does not necessarily assume causal association. This is incidence data, not prevalence data. (Ref. 31, p. 108)

341. C. This is a spot map and it does indicate where a physician is more likely to encounter Rocky Mountain spotted fever in his patients. These are not, however, rates; they represent incidence (a new event) rather than prevalence. (Ref. 31, p. 95)

342. D. During an epidemic of measles, isolation is impractical in the community at large. Closing schools is not recommended, but younger children should not be purposely exposed. Contacts need not be quarantined. Not all exposed children need be vaccinated, because they might already have resistance against the virus. (Ref. 21)

343. E. Early syphilis is considered transmittable, whereas late syphilis is rarely infectious. In contrast with the lesions of early syphilis, considerable destruction accompanies late lesions. Primary syphilis is usually detected by darkfield microscopy, whereas late syphilis is by serologic test and spinal fluid examination. (Ref. 22)

344. E. There is no risk to the fetus from passive immunization of pregnant women with immune globulin, nor is the fetus at risk when the mother is immunized with polio vaccine virus. Persons given measles, mumps, or rubella vaccine viruses do not transmit them, so these vaccines may be administered with safety to children of pregnant women. (Ref. 23)

345. E. All of the statements are correct. Since HIV infection persists in an individual, persons previously infected continue to remain at risk for developing AIDS. The incidence of AIDS continues to increase. In 1982, 747 cases were reported; in 1983, 2,124 cases were reported; in 1984, 4,569 cases were reported. And in 1985, 8,406 cases were reported. (Ref. 24)

346. A. In the United States, only about 10% to 15% of persons reported with hepatitis B infection had been identified as parenteral drug abusers. Nonetheless, control of hepatitis B among PDAs and persons in other risk groups is critical for limiting the spread of hepatitis B and delta agent in the general population. Many hepatitis B outbreaks have been demonstrated among PDAs over the past two decades and serologic surveys have demonstrated that HB virus infection is highly prevalent in this group. (Ref. 25)

347. A. Statements 1, 2, and 3 are correct. Leukopenia (cell counts less than 1500 cells/M3) is uncommon. Poor T-lymphocyte response to mitogen stimulation and altered humoral immunity are also found in children as compared with adults with AIDS. Because children who are infected with the AIDS virus have immunologic abnormalities associated therewith, some have questioned the advisability of immunizing them against the childhood infections. Immune deficient individuals have a higher risk of developing vaccine-associated poliomyelitis than normal individuals. Replication of live attenuated viruses may be enhanced in persons with immunodeficiency diseases and, theoretically, may produce serious adverse effects following immunization of symptomatic HIV infected patients. (Ref. 24)

348. E. Numerous occupational agents are associated with lung cancer, including chlorethers, chromates, polynuclear aromatic hydrocarbon compounds, as well as the four listed in the question. Tobacco smoke may interact with some of these agents, especially asbestos. (Ref. 7, p. 1251)

349. E. All of the statements are true. This question is given to emphasize the ODTS syndrome, which follows exposure to microbially contaminated vegetable dust. The syndrome is characterized by acute fever and respiratory symptoms, usually following exposure by 4 to 12 hours. General malaise, headache, and cough are common symptoms. ODTS is a self-limited illness, usually resolving within 24 hours. Preventive measures include storing vegetable matter in a way that limits microbial growth, and wearing appropriate respiratory protection when intense exposure to organic dusts cannot be avoided. (Ref. 26)

350. D. A first benefits period starts the first time one enters a hospital (after the hospital insurance begins). When a patient has been out of the hospital (or other facility primarily providing skilled nursing or rehabilitation services) for 60 days in a row, a new benefits period starts the next time one goes into a hospital. There is no limit to the number of benefits periods one can have. (Ref. 27).

References

1. Kuzma J: *Basic Statistics for the Health Sciences,* ed 1. Mountain View, CA, Mayfield Pub Co.

2. Nausner JS, Bahn K: *Epidemiology, Introductory Text,* Philadelphia, WB Saunders Co, 1985.

3. Levy BS, Wegman DH: *Occupational Health,* Boston, Little, Brown and Co, 1983.

4. Benenson AS: *Control of Communicable Disease in Man,* ed 14. Washington, DC, APHA, 1985.

5. US Department of Health and Human Services, Public Health Service: "Water-Related Disease Outbreaks Surveillance," Annual Summary 1980. Centers for Disease Control (CDC), Issued February 1982.

6. Advancedata, Vital and Health Statistics. The National Center for Health Statistics, No. 82, Issued June 16, 1982.

7. Last JM: *Preventive Medicine and Public Health,* ed 12. New York, Appleton-Century-Crofts, 1986.

8. Clark DW, Macmahon B: *Preventive Medicine,* ed 2. Boston, Little, Brown and Co, 1980.

9. Hanlon JJ, Pickett GE: *Public Health Administration and Practice,* ed 8. St Louis, Times Mirror/Mosby Co, 1984.

10. *CA—A Cancer Journal for Clinicians,* 35(1), Feb 1985.

11. Kahn HA: *An Introduction to Epidemiologic Methods,* New York, Oxford University Press, 1983.

12. *MMWR,* 31:31, Aug 13, 1982.

13. *MMWR,* 38:14, Apr 14, 1990.

14. *MMWR,* 39:3, Jan 26, 1990.

15. *MMWR,* 38:13, Apr 7, 1989.

16. *California Morbidity,* Feb 10, 1989.

17. *MMWR,* 34:22, June 7, 1985.

18. *MMWR,* 38:51, Dec 29, 1989.

19. *MMWR,* 34:23, June 14, 1985.

20. Advisory Memorandum #88 HHS, Apr 7, 1986.

21. *MMWR,* 38:S-9, Dec 29, 1989.

22. *MMWR,* 38:S-8, Sept 1, 1989.

23. *MMWR,* 33:S-1, Sept 28, 1984.

24. *MMWR,* 38:S-4, May 12, 1989.

25. *MMWR,* 39:52, Feb 9, 1990.

26. *MMWR,* 35:30, Aug 1, 1986.

27. The Superintendent of Documents: *Medicare Handbook, 1990.* Washington, DC, US Government Printing Office, 1990.

28. *Monthly Vital Statistics Report,* 34:13, Sept 1986.

29. Hanlon JJ, Pickett GE: *Public Health Administration and Practice,* ed 8. St Louis, CV Mosby Co, 1984.

30. Leavell HR, Clark EG: *Preventive Medicine for the Doctor in His Community,* ed 3. New York, McGraw-Hill Book Co, Inc, 1965.

31. *Morbidity and Mortality Weekly Report Annual Summary 1980. Centers for Disease Control, 29(54), Issued Sept 1981.*

32. *MMWR,* 26:15, Dec 16, 1977.

33. *MMWR,* 26:18, May 6, 1977.

34. *MMWR,* 34:4, Feb 1, 1985.

35. *MMWR,* 34:4, Feb 1, 1985.

36. *MMWR,* 39:RR-3, Mar 9, 1990.

37. US Institute of Medicine: *The Future of Public Health.* Washington, DC, National Academy Press, 1988.

38. A report of the Surgeon General: *Reducing the Health Consequences of Smoking, 25 Years of Progress.* US Department of Health and Human Services, 1989.

39. 1990 World Population Data Sheet. Washington, DC, Population Reference Bureau Inc.

4 Psychiatry

Carlyle H. Chan, MD, and Harry Prosen, MD

351. Bowlby described a sequential pattern of protest, despair, and detachment as a reaction to
   A. separation
   B. deprivation
   C. sensory bombardment
   D. narcotic withdrawal
   E. PCP intoxication

352. According to Erikson, the psychosocial task of the adolescent years is the achievement of
   A. basic trust; basic mistrust
   B. sense of ego identity; identity confusion
   C. generativity; stagnation
   D. integrity; despair
   E. sense of autonomy; pervasive sense of shame and doubt

353. Which of the following is **NOT** an ego function?
  A. Defense mechanisms
  B. Self-representation
  C. Reality testing
  D. Ability to tolerate frustration
  E. Instinctual drives

354. Which of the following is **NOT** characteristic of melancholia?
  A. Early morning awakening
  B. Psychomotor agitation
  C. Quality of mood different than mourning
  D. Mood worse in early morning
  E. Trouble falling asleep

355. The psychosocial task of the Oedipal period, according to Erikson, is the achievement of
  A. basic trust; basic mistrust
  B. autonomy; shame and doubt
  C. initiative; a sense of guilt
  D. industry; a sense of inferiority
  E. sense of ego identity; identity diffusion

356. Sensorimotor intelligence (Piaget) ends with the attainment of
  A. object constancy
  B. concrete operations
  C. formal operations
  D. rapprochement
  E. autism

357. The term "operant" refers to a class of responses that are
  A. elicited by some known stimulus
  B. emitted by the organism
  C. involuntary
  D. reflex
  E. operative in type S conditioning

358. In operant conditioning, negative reinforcement
  A. decreases the probability that a response will recur
  B. strengthens the response
  C. is usually ineffective
  D. is not contingent upon the response of the organism
  E. reinforces successively closer approximations to the desired behavior

359. The period defined in psychoanalysis as "latency" is roughly contemporaneous with the Piagetian stage of
  A. sensorimotor stage
  B. formal operations
  C. concrete operations
  D. preoperational thought
  E. logical thought

360. Which of the following is **NOT** typical of a psychomotorically retarded depressed patient?
  A. The patient sits with decreased facial expression
  B. The patient averts his eyes, looking downward in a fixed gaze
  C. The patient may speak and move slowly
  D. The patient tends to speak after a short, rapid inspiration
  E. There may be long pauses in the patient's speech

361. The following statements about the physician's feelings toward the patient are true **EXCEPT**
  A. the feelings may be based on the reality of the patient's appearance, character, and conflicts
  B. the feelings may be based on what the patient represents to the therapist's unconscious, in terms of early important figures
  C. strong negative feelings towards a patient will always interfere with treatment
  D. it is important to determine if negative feelings stem from reality or transferred aspects of the relationship
  E. most patients have some unlikable qualities

362. All of the following statements are accurate **EXCEPT**
- **A.** the superego is derived through identification with parents and their substitutes
- **B.** the superego observes and evaluates the self
- **C.** superego derivatives may be observed in phenomena metaphorically described as "an inner voice," and "inner authority," or "inner judge"
- **D.** the superego can be critical and punishing as well as praising and rewarding
- **E.** the superego includes all wishes that result from the perception and memories of gratification of basic physiological needs

363. A defense mechanism characteristic of the obsessive-compulsive disorder is
- **A.** acting-out
- **B.** projection
- **C.** undoing
- **D.** conversion
- **E.** denial

364. Splitting is
- **A.** primitive idealization
- **B.** projective identification
- **C.** division of the self and external objects into "all good" and "all bad"
- **D.** ambivalence
- **E.** none of the above

365. After learning to press a lever to obtain food, a chicken is subsequently trained to press the lever 15 times for each pellet. The principle of learning which has **NOT** been demonstrated is
- **A.** partial reinforcement
- **B.** fixed-ratio scheduling
- **C.** operant conditioning
- **D.** classical conditioning
- **E.** instrumental conditioning

366. A "script" in transactional analysis refers to
  A. basic existential position
  B. important early life decisions
  C. early nonverbal messages
  D. rackets
  E. stroking patterns

367. "Primary process" is a psychoanalytic term used to describe a certain kind of thinking which is **NOT**
  A. primitive, dominated by emotions
  B. characteristic of the infant
  C. prominent in dreams and psychoses
  D. taking into consideration external reality
  E. typical of psychotic states

368. "Primary gain" refers to
  A. impulse control
  B. control of anxiety
  C. gratification of warded-off wish
  D. avoidance of unpleasant demands
  E. B and C

369. The most common extrapyramidal side effect to develop within the first few days of beginning with a high potency major tranquilizer is
  A. dystonia
  B. akinesia
  C. tardive dyskinesia
  D. rabbit syndrome
  E. Parkinson syndrome

370. Defense mechanisms
  A. are unconscious
  B. are not ego functions
  C. are not used by normal adults
  D. develop after the resolution of the Oedipal conflict
  E. include suppression

371. Organic delusional syndrome is characterized by
A. confabulation
B. prominent hallucinations
C. ability to maintain attention to external stimuli
D. no evidence of organic etiology
E. loss of intellectual abilities

372. Alcohol hallucinosis is characterized by all of the following **EXCEPT**
A. high incidence in people in their early 20s
B. voices discussing patient in the third person
C. auditory hallucinations
D. duration of many months
E. occurrence within a few days of alcohol withdrawal

373. Which of the following is **NOT** true of Alzheimer's disease?
A. Marked atrophy of basal ganglia
B. No gross neurological deficits, except possible aphasia
C. Generalized cortical atrophy
D. Neurofibrillary tangles
E. Misdiagnosis may run as high as 50%

374. Management of tricyclic overdose may include all of the following **EXCEPT**
A. gastric aspiration
B. alkalinization
C. EKG monitoring
D. ventilatory assistance
E. naloxone (Narcan) 0.4 mg/IV

375. According to Kinsey, what percentage of males are predominantly homosexual in orientation for at least three years between ages 16 and 55?
A. 5%
B. 1%
C. 13%
D. 32%
E. 28%

376. Which of the following is **NOT** characteristic of the schizotypal personality disorder?
   A. Magical thinking
   B. Relatives with schizophrenic disorder
   C. Illusions
   D. Odd speech
   E. Thought insertion or withdrawal

377. The borderline personality disorder is characterized by all of the following **EXCEPT**
   A. unstable, intense relationships
   B. inappropriate intense anger
   C. potentially self-damaging impulsiveness
   D. social withdrawal
   E. disturbances in identity

378. Which of the following is **NOT** characteristic of posttraumatic stress disorder?
   A. Numbing of responsiveness
   B. Increased arousal
   C. Impairment of memory of trauma
   D. Thought insertion
   E. Reexperiencing of past trauma

379. A 25-year-old man has a six-month history of believing that "people" are trying to put "bad thoughts" into his head and to make him do "bad things." He finds special messages from them in TV news reports and in the newspaper. He is agitated, pacing constantly. His affect is inappropriate, and he laughs as he tells of his persecution. Physical examination is within normal limits. Drug screen is negative. The pharmacologic treatment of choice is
   A. clorazepate (Tranxene) 30 mg/d
   B. lorazepam (Ativan) 20 mg/d
   C. haloperidol (Haldol) 5 to 20 mg/d
   D. amitryptiline (Elavil) 150 mg/d
   E. imipramine (Tofranil) 25 mg/d

380. Which of the following statements is **NOT** true?
- **A.** Confidentiality is the obligation of the professional to keep in confidence (ie, from third parties) whatever information is shared by the patient absent permission
- **B.** Privilege is the right of patients to ban or exclude from judicial or quasi judicial settings testimony about material that has been revealed within the professional relationship
- **C.** Breaching confidentiality may be a liability issue
- **D.** Privilege belongs to the doctor, who may decide to exercise it or not
- **E.** Tarasoff holds that a therapist has a duty to warn intended victims

381. The following are all criteria for establishing testamentary capacity **EXCEPT**
- **A.** knowing that one is making a will
- **B.** knowing the objects of one's bounty
- **C.** knowing the extent of one's property
- **D.** no undue influence
- **E.** knowing the difference between right and wrong

382. The hypothesis that a depressed person may maintain his depression by how he thinks is a tenet of
- **A.** Heinz Kohut
- **B.** Aaron Beck
- **C.** Margaret Mahler
- **D.** Otto Kernberg
- **E.** Fritz Perls

383. The WISC
- **A.** is a projective test
- **B.** includes only verbal tests
- **C.** tests only innate intellectual ability
- **D.** includes only nonverbal tests
- **E.** is the child equivalent of the WAIS

384. Which of the following is a piperidine phenothiazine?
A. Haloperidol
B. Stelazine
C. Mellaril
D. Trilafon
E. Prolixin

385. Chlordiazepoxide exerts its therapeutic effects most quickly when it is administered
A. orally
B. intramuscularly
C. subcutaneously
D. with anticholinergics
E. with stimulants

386. A definitive contraindication to electroconvulsive therapy (ECT) is
A. brain tumor
B. threatened retinal detachment
C. aortic aneurysm
D. myocardial disease
E. glaucoma

387. Thioridazine, in high doses, is especially likely to produce
A. retinitis pigmentosa
B. cholestatic jaundice
C. orthostatic hypotension
D. tardive dyskinesia
E. akinesia

388. The following are early symptoms of lithium toxicity EXCEPT
A. heightened auditory sensitivity
B. dysarthria
C. ataxia
D. coarse tremor
E. neuromuscular irritability

389. Which of the following drugs will tricyclic antidepressants block?
   A. Haldol
   B. Griseofulvin
   C. Reserpine
   D. Propranolol
   E. Guanethidine

390. Which of the following has the **LEAST** anticholinergic side effects?
   A. Desipramine
   B. Amitryptiline
   C. Nortriptyline
   D. Trazodone
   E. Imipramine

391. Which of the following statements is **NOT** true?
   A. There is a disproportionate number of schizophrenic patients in lower socioeconomic classes in industrialized nations
   B. The lifetime prevalence for schizophrenia in the United States is about 6.5%
   C. Part of the homeless problem in large cities may be related to inadequate follow-up of deinstitutionalized schizophrenic patients
   D. In the northern hemisphere, more schizophrenic patients are born in the winter months of January to April
   E. Approximately 90% of patients in treatment for schizophrenia are between 15 and 54 years of age

392. Which of the following is **NOT** considered an example of primary prevention?
   A. Genetic counseling of persons with a family history of PKU (phenylketonuria)
   B. Intensive intervention with blind infants
   C. Competency training in preschool settings
   D. Widow-to-widow self-help groups
   E. Lithium clinics

393. True statements concerning grief and bereavement include
  A. initial physical symptoms that may be indistinguishable from depression
  B. widowers and widows have an increased death rate in the first year following the death of a spouse
  C. identification phenomenon may occur
  D. grief may extend from 1 to 2 years
  E. all of the above

394. The mental-status examination includes all the following EXCEPT
  A. general appearance, manner, and attitude
  B. affect, thought processes
  C. memory
  D. judgment, insight
  E. personal history

395. In the Midtown Manhattan Study, what percentage of questionnaire respondents were considered "less than well"?
  A. 10%
  B. 24%
  C. 23.4%
  D. 81.5%
  E. 87%

396. Causes of mental retardation include all of the following EXCEPT
  A. too much early stimulation
  B. hereditary disorders
  C. pregnancy problems and premarital morbidity
  D. acquired childhood diseases
  E. unknown causes

397. A 26-year-old woman suffers from panic attacks—three or four a week. The most appropriate pharmacological choice would be
A. imipramine (Tofranil) 25 mg/d, increasing the dose gradually if necessary
B. imipramine (Tofranil) 150 to 300 mg/d
C. oxazepam (Serax) 30 mg/d
D. clorazepate (Tranxene) 30 mg/d
E. chlordiazepoxide (Librium) 100 mg/d

398. Anticholinergic side effects include all of the following EXCEPT
A. peripheral vasodilation
B. delirium
C. dry mucous membranes
D. impaired visual accommodation
E. peripheral neuropathy

399. You admit to the hospital a patient with a six-month history of marked thought disorder, inappropriate affect, and auditory hallucinations. His level of consciousness is clear, his orientation and memory good. Making a diagnosis of schizophrenic disorder, you prescribe trifluoperazine (Stelazine) 5 mg/qid. The next day he develops a painful spasm of the sternocleidomastoid muscle, which twists his head to the right. He has developed
A. muscular dystonia
B. tardive dyskinesia
C. akathesia
D. akinesia
E. parkinsonism

400. The affect in schizophrenia is NOT
A. blunted
B. flat
C. inappropriate
D. infectious
E. restricted

401. Anorexia nervosa is characterized by all the following **EXCEPT**
  A. intense fear of becoming obese
  B. weight loss of at least 25%
  C. disturbance of body image
  D. onset in early latency years
  E. delusional perception

402. The following statements about the differential diagnosis of Avoidant Disorder of Childhood are true **EXCEPT**
  A. socially reticent children are slow to warm up, but can respond after a short time with no impaired peer interaction
  B. in Separation Anxiety Disorder, anxiety is focused on separation from the home or significant attachment figures rather than unfamiliar persons per se
  C. in Adjustment Disorder with Withdrawal, withdrawal is related to past psychosocial stressors, and lasts at least six months
  D. in Overanxious Disorder, anxiety is not focused on contact with unfamiliar people
  E. the disturbance in Avoidant Disorder of Childhood is not sufficiently pervasive and persistent to warrant the diagnosis of Avoidant Personality Disorder

403. Diagnostic criteria for Conduct Disorder include all of the following **EXCEPT**
  A. has deliberately engaged in fire setting
  B. often lies (other than to avoid physical or sexual abuse)
  C. conduct disturbance lasting at least three weeks
  D. has been physically cruel to animals
  E. has broken into someone else's house, building, or car

404. Dementia is characterized by
  A. loss of intellectual ability
  B. memory impairment
  C. ability to attend to external stimuli
  D. disturbances of higher cortical functions
  E. disturbances in memory

405. Alcohol-withdrawal delirium is characterized by
   A. onset usually within one week of abstinence
   B. transient hallucinations or illusions
   C. marked autonomic hyperactivity
   D. reduced ability to maintain attention to external stimuli
   E. all of the above

406. Which of the following is **NOT** true with regard to informed consent?
   A. To perform any procedure or any touching of a patient in a medical center without consent constitutes battery
   B. The consent process contains three elements: information, competence, and voluntariness
   C. It requires the physician to relate sufficient information to the patient to allow him or her to decide if the procedure is acceptable in light of its risks and benefits
   D. It is never added to the claim of malpractice
   E. Consent forms must not substitute for a physician-patient dialogue

407. In negligence,
   A. a standard of care must exist
   B. a duty must be owed by the defendant or someone for whose conduct he is answerable
   C. the duty must be owed the plaintiff
   D. there must be a breach of the duty
   E. all of the above

408. The following statements concerning diabetes mellitus are true **EXCEPT**
- **A.** inadequate patient compliance with prescribed treatment is the most common cause for psychiatric consultations
- **B.** personality type may determine compliance with treatment regimens
- **C.** positive reinforcement is an important aid to developing new behaviors regarding diet, urine checks, and insulin administration
- **D.** special caution must be used in adjusting insulin levels when ECT is used for treating depression in a patient with diabetes
- **E.** family support, continuity of care, and trust in the treating physician all contribute to successful treatment

409. The diagnostic criteria for psychological factors affecting physical conditions include all the following **EXCEPT**
- **A.** temporal relationship between meaningful stimulus and initiation or exacerbation of condition
- **B.** demonstrable organic pathology
- **C.** known pathophysiology
- **D.** absence of somatoform disorder
- **E.** symbolism

410. The best predictor of obesity in the United States is
- **A.** type-A personality
- **B.** low socioeconomic class
- **C.** life change
- **D.** learned helplessness
- **E.** heredity

411. Neuroimaging techniques include the following **EXCEPT**
   A. magnetic resonance
   B. positron emission tomography
   C. computed tomography
   D. electromyogram
   E. cerebral angiography

412. A patient with a clearly acute manic presentation of bipolar disorder does not respond to lithium carbonate. Treatment alternatives include all of the following **EXCEPT**
   A. carbamazapine (Tegretol)
   B. valproic acid
   C. thyroid potentiation
   D. clonazepan
   E. antipsychotics

413. The following statements are true about prevention in community mental health **EXCEPT**
   A. primary prevention strives to decrease the incidence of mental disorders
   B. secondary prevention involves early case finding and treatment to minimize duration and avoid permanent disability with an aim to decrease prevalence
   C. primary prevention of major mental illness through early childhood intervention has been highly successful
   D. tertiary prevention's goal is to reduce the prevalence of residual defects or disabilities through rehabilitation
   E. care of chronically mentally ill may involve both goals of tertiary treatment

414. Forms of Brief Dynamic Psychotherapy include all of the following **EXCEPT**
   A. broad-focused short-term dynamic
   B. cognitive
   C. time limited
   D. short-term anxiety provoking
   E. short-term interpersonal

DIRECTIONS (Questions 415–423): This section consists of clinical situations, each followed by a series of questions. Study each situation and select the **one** best answer to each question following it.

Question 415: A 30-year-old resident is noted to be especially frugal, rigid, and punctual. In most areas of his life he is meticulous, but his desk and bedroom are very messy. He tends to be obsequious with superiors and rather sadistic with the medical students in his control. Occasionally he has temper tantrums.

415. From a psychodynamic point of view, these characteristics are derivatives of which of the following stages of development?
   A. Oral
   B. Anal
   C. Phallic
   D. Oedipal
   E. Latency

Question 416: A 10-year-old boy confuses letters and words as he writes and reads, mixes "b" with "d" and "p" with "g." He is intelligent, but becomes very frustrated at school.

416. The most likely diagnosis is
   A. school phobia
   B. brain tumor
   C. pervasive developmental disorder
   D. gender-identity disorder
   E. developmental reading disorder

Questions 417–420: A 38-year-old, white, married woman comes to her family physician with a history of vague abdominal pains. She is certain she has cancer. Exhaustive medical examinations and general hospitalizations have failed to reveal any abnormality other than "spastic colitis." Yet, she continues to believe she has cancer, but "The doctors just haven't found it yet." She wakes up early in the morning (about 4 am). She has lost at least 15 pounds in the last six weeks (a fact she attributes to cancer). Her speech is monotonous and slow. Tears come to her eyes as she begins to talk about the fact that her youngest child joined the Navy five months ago. Since then she has felt useless, and has found no pleasure in anything. Although she never feels good, she believes she feels worse in the morning. She previously had been well. She denies any previous history of similar symptoms. She has received no prior psychiatric help.

417. Which of the following is the most likely axis I diagnosis?
      A. Hypochondriasis
      B. Major depression, single episode
      C. Bipolar disorder
      D. Cancer of the pancreas
      E. Somatoform disorder

418. Which of the following is the most likely axis II diagnosis?
      A. Compulsive personality disorder
      B. No diagnosis
      C. Diagnosis deferred
      D. 3
      E. 6

419. The woman
      A. should not be asked about suicide
      B. should be asked about suicide
      C. may make a suicidal manipulative gesture
      D. is very unlikely to attempt suicide
      E. none of the above

420. Which of the following would allow the therapist to be less concerned about suicide?
   A. Family history of suicide
   B. Patient's mood improves
   C. Patient tells about her suicidal ideas
   D. Patient made a suicidal gesture two weeks ago
   E. None of the above

Questions 421–422: A 40-year-old woman has a several-year history of multiple somatic complaints for which she has been "worked up" and treated by many competent physicians. However, all their efforts have failed to influence her chronic but fluctuating history of malaise and somatic distress. She now complains of an array of symptoms: dysparenuria, irregular and painful menses, shortness of breath, light-headedness, nausea, heartburn, frequency of urination, "weak spells," and "twitching legs," in a vague but dramatic manner. Physician examination and laboratory testing again disclose no abnormality.

421. The most likely diagnosis is
   A. Munchausen's syndrome
   B. factitious disorder
   C. malingering
   D. hypochondriasis
   E. somatization disorder

422. Which of the following is **NOT** correct?
   A. Brain norepinephrine is metabolized to 3,4 MHPG
   B. Norpramin blocks the reuptake of norepinephrine
   C. Amitryptiline blocks the reuptake of only norepinephrine
   D. Some cases of depression are associated with low brain levels of norepinephrine
   E. Urinary levels of 3,4 MHPG are low in some groups of depressed people

Question 423: A young woman who is being treated with medication for an atypical depression goes to a party where she eats chicken-liver pate and cheese. She develops a severe headache. Her physical examination is normal, except for a blood pressure reading of 200/130 mmHg.

423. The woman probably has been taking
    A. a butyrophenone
    B. a phenothiazine
    C. a monoamine-oxidase inhibitor
    D. a tricyclic antidepressant
    E. lithium carbonate

DIRECTIONS (Questions 424–433): Each group of questions below consists of lettered headings followed by a list of numbered words or statements. For each numbered word or statement, select the **one** lettered heading that is most closely associated with it. Each lettered heading may be selected once, more than once, or not at all.

Questions 424–427:
    A. Repression
    B. Denial
    C. Displacement
    D. Projection
    E. Reaction-formation
    F. Dissociation

424. Involuntary banishment of feelings, ideas, or impulses from awareness

425. Consciously intolerable facts or thoughts are ignored

426. Attribution of one's own wishes or attitudes to another

427. Feeling or impulse originally directed toward one person is transferred to another

Questions 428–430:
  A. "Bonnie Rule"
  B. Durham decision
  C. M'Naghten

428. Judge David Bazelon

429. Attempted assassination of Sir Robert Peel

430. Incorporated into the Insanity Defense Reform Act of 1985

Questions 431–433:
  A. Increases REM sleep
  B. Decreases REM sleep
  C. No effect on REM sleep

431. Barbiturates

432. Benzodiazepines

433. Reserpine

DIRECTIONS (Questions 434–444): Each set of lettered headings below is followed by a list of numbered words or phrases. For each numbered word or phrase select
  A if the item is associated with **A** only,
  B if the item is associated with **B** only,
  C if the item is associated with both **A** and **B**,
  D if the item is associated with neither **A** nor **B**.

Questions 434–439:
  A. Psychoanalytic understanding of borderline personality
  B. Psychoanalytic understanding of narcissistic personality
  C. Both
  D. Neither

434. Lack of cohesive sense of self

435. Splitting

436. Preservation of reality testing

437. Self and object representations divided into "all good" and "all bad"

438. Use of "self-objects"

439. Genetically related to schizophrenic disorder

Questions 440–442:
  A. Loss of intellectual functioning
  B. Fluctuating level of consciousness
  C. Both
  D. Neither

440. Delirium

441. Dementia

442. Hallucinosis

Questions 443–444:
  A. Non-REM sleep
  B. REM sleep
  C. Both
  D. Neither

443. Young adults average 7 to 8 hours sleep per night with six hours in this type of sleep

444. Newborn children average 16 to 18 hours of sleep with roughly 8 to 9 hours in this type of sleep

DIRECTIONS (Questions 445–466): For each of the questions or incomplete statements below, **one** or **more** of the answers or completions given is correct. Select

A   if only **1, 2,** and **3** are correct,
B   if only **1** and **3** are correct,
C   if only **2** and **4** are correct,
D   if only **4** is correct,
E   if **all** are correct.

445. Middle-adult years may be characterized by
    1. a period of surrendering illusions
    2. emotionally processing the issue of existence
    3. an imperative to act
    4. achievement of integrity

446. The "pre-psychologic" phase of attachment (first two months) is characterized by
    1. following moving objects with eyes
    2. demonstrating reflex grimace without eye crinkling
    3. listening
    4. preferring mother's voice by one month of age

447. A 22-year-old woman is seen by her family practitioner who notes that she has been depressed for the past month and complains of trouble falling asleep, fatigue, a 10-pound weight gain, decreased concentration, and feeling slowed down. The physician gives her a month's supply of amitryptiline (Elavil) 50mg/hs and makes an appointment for the following month. Which of the following is(are) correct?
    1. The physician may not have given her an antidepressant dose of Elavil
    2. The physician did not consider possible additive effects of psychotherapy in the treatment plan
    3. The physician gave her a lethal supply of the drug
    4. The physician should have begun with lithium rather than a tricyclic

Directions Summarized				
A	**B**	**C**	**D**	**E**
1,2,3	*1,3*	*2,4*	*4*	*All* are
only	only	only	only	correct

448. In a study of a coronary care unit, the most frequent reasons for consultation were
1. depression
2. management of behavior
3. anxiety
4. staff burn-out

449. Which of the following drugs are not to be used to treat psychotic depression?
1. Chlordiazepoxide (Librium)
2. Desipramine (Norpramin)
3. Nortriptyline (Pamelor)
4. Loxapine (Loxitane)

450. Which of the following are projective tests?
1. TAT
2. CAT
3. Rorschach
4. MMPI

451. Which of the following are tests for assessing brain damage?
1. Halstead-Reitan
2. Luria-Nebraska
3. Boston process neuropsychological approach
4. Bender-Gestalt

452. Tardive dyskinesia
1. may worsen with psychotic medication discontinuation
2. may be suppressed with higher doses of the antipsychotic medication
3. may fade after an initial worsening
4. may be mistaken for acute choreoathetotic reactions

453. The piperazine group of phenothiazines tend to produce more
1. postural hypotension
2. anticholinergic effects
3. sedation
4. extrapyramidal symptoms

454. If a patient who is taking disulfiram daily begins to drink, he is likely to experience
1. throbbing headaches
2. nausea and vomiting
3. palpitations
4. diaphoresis

455. Indolamines include
1. norepinephrine
2. epinephrine
3. dopamine
4. 5-OH tryptamine

456. Antidepressants are quite effective in the treatment of
1. bulemia nervosa
2. chronic pain syndrome
3. panic disorder
4. schizophrenia, residual type

457. The MMPI
1. is an objective questionnaire
2. is a projective test
3. includes a scale to detect simulation
4. is a means to tap into unconscious fantasy

Directions Summarized				
A	**B**	**C**	**D**	**E**
1,2,3	*1,3*	*2,4*	*4*	*All* are
only	only	only	only	correct

458. Which of the following are true statements concerning DSM III-R?
  1. Each mental disorder is conceptualized as a clinically significant behavioral or psychological syndrome or pattern
  2. The principal diagnosis must be an axis I diagnosis
  3. Stressors are rated in terms of the severity an "average" person would experience
  4. Disturbances limited to the relationship between the individual and society are conceptualized as mental disorders

459. Which of the following is(are) true of delirium?
  1. It rarely persists over one month
  2. Fluctuations in symptoms are common
  3. Psychomotor activity is disturbed
  4. It is most common in middle age

460. A delusion is an idea that is
  1. incorrect
  2. not correctable by facts to the contrary
  3. not held by the majority of one's peers
  4. always a symptom of schizophrenic disorder

461. In evaluating a person's competency to make a will, the physician should appropriately investigate his knowledge
  1. that he is making a will
  2. of the nature and extent of his property
  3. of the natural objects of his bounty
  4. of right and wrong

462. Which of the following elements is(are) important in the American Law Institute's test of criminal responsibility?
1. Mental disease or defect
2. Lack of substantial capacity
3. Appreciation
4. Conformity of conduct to requirements of law

463. If a depressed patient fails to respond to a tricyclic antidepressant, acceptable changes to the treatment plan include
1. increasing the dosage
2. checking the serum level of the medication (if one is obtainable)
3. L-triodothyronine (T-3) supplementation
4. lithium augmentation

464. Prior to initiating lithium carbonate therapy, the following laboratory tests should be conducted.
1. Thyroid function studies (T4, T3RU, FT41, TSH)
2. CBC
3. Serum creatinine level
4. Electrolyte screen

465. Patients with borderline personality disorder
1. can show inappropriate intense anger
2. may exhibit self-mutilative behavior
3. report chronic feelings of emptiness or boredom
4. display perfectionism that interferes with task completion

466. The following statements are true about a light treatment for seasonal affective disorder.
1. Bright light (2500 lux) is needed
2. May only be required in the morning
3. Patients may respond after only 2 to 4 days of treatment
4. Requires 2 to 4 hours of continuous staring directly into the light source

Explanatory Answers

351. A. Bowlby described the sequential pattern of protest, despair, and detachment as a reaction to separation from their mother in children between six months and three years of age. (Ref. 1, p. 84)

352. B. The major task of adolescence is the attainment of a sense of identity. Erik Erikson's schema of eight sequential psychosocial tasks has been a major contribution to our understanding. Erikson's psychosocial tasks should be distinguished from the traditional psychoanalytic stages of psychosexual development: oral, anal, phallic, Oedipal, etc. (Ref. 2, pp. 406–408)

353. E. The ego is traditionally defined by its functions. These include the abilities to tolerate stress, tolerate anxiety, delay gratification, and distinguish reality from fantasy ("reality testing"), as well as internal representations of oneself and others, interpersonal relationships ("object relationships"), the defense mechanisms, and the "autonomous functions" of perception, memory, and mobility. Drives are the psychological representatives of the instincts. (Ref. 1, pp. 140–142)

354. E. Trouble falling asleep is not typical of melancholia, in contrast to early morning awakening. Patients with symptoms of melancholia seem more likely to respond to tricyclic medications. (Ref. 5, pp. 223–224)

355. C. The major psychosocial task of the Oedipal period is the development of a sense of initiative. (Ref. 1, p. 39)

356. A. For Piaget, intelligence is biological. The first stage (the stage of sensorimotor intelligence) begins with the reflex arc and ends with the attainment of object constancy. That is, the infant (about 18 months) knows an object continues to exist even if he cannot see it. (Ref. 1, p. 82)

357. B. Operant responses are frequently referred to as voluntary, as opposed to involuntary (or reflex), behavior. (Ref. 1, p. 86)

358. A. A negative reinforcement is an event likely to decrease the probability of a response's recurrence following the removal of a negative stimulus. (Ref. 1, p. 87)

359. C. The phase of concrete operations generally is achieved during the latency stage, supplanting the intuitive phase (4 to 7 years), during which concepts of space and time are acquired. (Ref. 1, p. 82)

360. D. Typically, patients who are depressed tend to speak after expiration. Speaking after a short, rapid inspiration is more typical of people who are anxious. (Ref. 4, p. 311)

361. C. The tendency for physicians to displace feelings from earlier figures in their life onto the patient is called countertransference. Understanding countertransference feelings may help avoid interference in proceeding with treatment. (Ref. 4, p. 20)

362. E. The description in E is that of the id. (Ref. 6, pp. 90, 189–190)

363. C. Undoing is a compulsive act performed in an attempt to undo or prevent the anticipated consequence of a frightening thought or impulse. The defenses used typically in obsessive-compulsive disorder are: reaction-formation, isolation, undoing, intellectualization, and displacement. (Ref. 6, p. 132)

364. C. Borderline and psychotic conditions are typified by the predominance of primitive defenses such as splitting. Splitting refers to the process of experiencing the self or others as "all good" or "all bad," rather than having both good and bad properties. (Ref. 2, p. 375)

365. D. Instrumental or operant conditioning is a form of learning in which relatively spontaneous behavior is punished or rewarded. Partial reinforcement consists of different forms of intermittent reward of which fixed-ratio scheduling is one. In fixed-ratio scheduling, a specific number of responses must occur before a reward is given. In classical or Pavlovian conditioning, in contrast, a stimulus which once had no ability to bring on a specific response becomes able to do so. (Ref. 1, pp. 86–87)

174 / Clinical Sciences

366. B. In transactional analysis, a script is determined by life decisions that are lodged in the child ego state as a result of repetitive or especially strong parental injunctions, but they can be altered by redesign. (Ref. 2, p. 430)

367. D. Primary process is the illogical mode of thinking which is typical of the unconscious. It is the type of thinking found in dreams and in psychosis. Secondary process, in contrast, refers to orderly mental activity which gives due regard to everyday logic. (Ref. 6, pp. 148–149)

368. E. Primary gain should not be confused with secondary gain—external rewards (such as attention and financial settlements or avoidance of duties). Primary gain is manifest in control of anxiety and gratification of warded-off wishes. (Ref. 2, p. 104)

369. A. Muscular dystonias are most likely to develop within the first 48 hours after beginning a major tranquilizer. (Ref. 7, p. 414)

370. A. Defense mechanisms are unconscious. They function automatically, out of our awareness, to reduce conflict and anxiety. Coping mechanisms include both the defense mechanisms and conscious mechanisms such as suppression and fact-finding. (Ref. 6, pp. 48–49)

371. C. The organic delusional syndrome is characterized by clear consciousness and delusions. Dementia is characterized by loss of intellectual abilities in a state of clear consciousness. In delirium, the state of consciousness fluctuates. (Ref. 5, p. 109)

372. A. The typical age for the onset of alcohol hallucinosis is about 40, after 10 or more years of heavy drinking. Like delirium tremens, alcohol hallucinosis is a withdrawal phenomenon. There is, however, a clear state of consciousness; the patient has auditory, as opposed to visual and tactile, hallucinations. (Ref. 5, p. 132)

373. A. Alzheimer's disease is the most common of the presenile dementias. Pathologically, we cannot currently distinguish it from senile dementia. Histologically, it is characterized by diffuse neuronal loss, neurofibrillary tangles, especially in the hippocampal gyrus, and senile plaques in the grey matter. There is no destruction of the basal ganglia, as occurs in Huntington's chorea, and Jakob-Creutzfeldt's disease. The student should distinguish the reversible forms of dementia, such as vitamin deficiency (eg, B_{12}), normal pressure hydrocephalus, and the pseudodementia of depression in the elderly. (Ref. 3, p. 614)

374. E. Tricyclic overdose is now one of the most common causes of poisoning. Narcan is a narcotic antagonist. (Ref. 7, pp. 438–439)

375. C. Although 30 years old, the Kinsey data remains one of the best sources of information about the prevalence of homosexual behavior in American society. Four percent of men were reported to be exclusively homosexual throughout their adult life. Thirteen percent were predominantly homosexual for at least 3 years, and another 13% reported having reacted erotically to males after childhood. The student should distinguish homosexual behavior from transvestism and transsexualism. (Ref. 2, p. 1087)

376. E. Thought insertion or withdrawal is one of the A criteria for schizophrenic disorders in the DSM III-R classification. (Ref. 5, pp. 341–342)

377. D. The borderline personality is characterized by general instability of mood, identity, and interpersonal relationships. Psychodynamically, these people use much denial and splitting, experiencing themselves and others as either "all good" or "all bad." Developmentally, they seem fixated at the "rapprochement stage" of separation-individuation (Mahler). (Ref. 5, p. 347)

378. D. Thought insertion is a Schneiderian criterion for schizophrenia and was chosen as one of the diagnostic criteria for schizophrenic disorder in DSM III. (Ref. 1, p. 503)

379. C. The most likely diagnosis for this man is schizophrenic disorder. Consequently, a major tranquilizer is the treatment of choice. (Ref. 2, pp. 770–765, 778–780)

380. D. The belief that privilege somehow belongs to the doctor is one of the most widely held misconceptions. (Ref. 3, pp. 2118–2120)

381. E. Testamentary capacity refers to the capacity to make a will. Knowledge as to whether the particular act was right or wrong is one of the M'Naghten criteria for insanity. The student should note that terms such as "insanity" and "competence" have specific legal definitions. These are not medical terms. (Ref. 3, pp. 2114, 2121)

382. B. Aaron Beck's work on the cognitive therapy of depression is now quite accepted. A similar approach can be found in transactional analysis (see Allen, Allen *Psychiatry: A Guide*, New York, Medical Examination Publishing Company, 1984) and in Albert Ellis' Rational Emotive Therapy. However, these ideas go back to the Greek philosopher Heraclitus of Ephesus. Heinz Kohut is most famous for his work on narcissism and disorders of the self. Kernberg has brought object-relations theory into the mainstream of modern psychiatry. Margaret Mahler's work on the separation-individuation stage of child development is already classic. These three have revitalized psychoanalysis. Fritz Perls developed gestalt therapy. (Ref. 4, p. 1542)

383. E. The WISC (Wechsler Intelligence Scale for Children) is the child version of the WAIS. Like the WAIS, it includes verbal and performance tests. (Ref. 1, p. 410)

384. C. Phenothiazines can be subdivided chemically into three groups: aliphatics, piperazines, and piperidines. All these groups, like nonphenothiazine antipsychotics such as butyrophenones, share one property in common: they block dopamine receptors. (Ref. 7, p. 447)

385. A. Chlordiazepoxide (Librium) is poorly absorbed when given intramuscularly. (Ref. 7, p. 2323)

386. A. There are almost no absolute contraindications for ECT, given under modern anesthesia and with a modern machine. A brain tumor is one contraindication, however, because of ECT-associated breakdown of the blood-brain barrier and the sudden increase in intracranial pressure during a seizure. Intraocular pressure decreases, however, so glaucoma is not a contraindication. It should be noted that with modern methods, ECT is really safer than the use of tricyclic medication. (Ref. 3, p. 1677)

387. A. Thioridazine (Mellaril) may cause retinitis pigmentosa if doses exceed 800 mg/d. (Ref. 7, p. 413)

388. A. Early symptoms of toxicity include dysarthria, ataxia, coarse tremor, and neuromuscular irritability. Patients in acute mania can usually tolerate quite high doses of lithium, but the dosage must be rapidly reduced to maintenance levels when the acute attack has subsided. More severe intoxication may present as delirium, myoclonus, impaired consciousness, seizures, coma, and death. (Ref. 3, p. 1660)

389. E. The blockage of a rather nonspecific transport system by the tricyclics will block the access of guanethidine to its site of action in the sympathetic nerve terminals. (Ref. 7, p. 438)

390. A. Trazodone (Desyrel) has the weakest anticholinergic properties. Anticholinergic side effects include: palpitations, loss of visual accommodation, dry mouth, aggravation of narrow-angle glaucoma, urinary retention, and paralytic ileus, as well as delirium. (Ref. 3, p. 1643)

391. B. The actual lifetime prevalence is 1%. (Ref. 1, pp. 254–255)

392. E. Primary prevention is prevention in the layman's use of the word. Secondary prevention is early diagnosis and treatment. (Ref. 3, pp. 2067–2068)

393. E. Grieving is a dynamic state with symptoms lasting until the person has had the opportunity to experience the entire calendar year at least once without the lost person. The bereaved person may take on the qualities, mannerisms, or characteristics of the deceased to perpetuate that person in some way. (Ref. 1, pp. 53–54)

394. E. Personal history is not part of the mental-status examination. Indeed, the mental-status examination is a description of a person's mental state at one point in time—much like the "still pictures" advertising a movie. The mental-status does not tell where the person has been or where he is going. (Ref. 1, pp. 157–158)

395. D. (Ref. 1, p. 107)

396. A. (Ref. 3, pp. 1739–1746)

397. A. Imipramine (Tofranil) starting at a low dose can prevent panic attacks. Some patients require antidepressant doses. MAOIs and high-potency benzodiazepines have also shown efficacy. New evidence shows that patients may respond to low and moderate dosages of benzodiazepines. To treat any anticipatory anxiety, however, such as the fear of having a panic attack, an antianxiety agent will probably be necessary. (Ref. 2, pp. 970–971)

398. E. The peripheral syndrome is characterized by mydriasis, dry mouth, cycloplegia, flushing, and tachycardia. Sometimes, however, sweating may occur with the tricyclic antidepressants. (Ref. 8, p. 254)

399. A. Muscular dystonia is the most common extrapyramidal side effect within the first few days after beginning treatment with a major tranquilizer, accounting for about 6% of extrapyramidal symptoms. Akathesia, however, is the most common side effect overall. (Ref. 7, p. 414)

400. D. An infectious affect is more characteristic of mania. (Ref. 5, p. 195)

401. D. Although it may occur in the latency years, anorexia nervosa typically occurs during adolescence. (Ref. 5, p. 66)

402. C. The true criteria in Adjustment Disorder with Withdrawal is withdrawal related to a recent psychosocial stressor and is less than six months. (Ref. 5, pp. 62–63)

403. C. A disturbance of conduct must last at least six months. There are a total of 13 behaviors, of which at least three must be present. (Ref. 5, p. 55)

404. C. Loss of intellectual abilities is the essential feature of dementia. Memory, judgment, abstract thought, and other higher cortical functions decline. The diagnosis is not made if these features are due to a reduced ability to maintain or shift attention to external stimuli, as in delirium. (Ref. 5, p. 103)

405. E. (Ref. 5, pp. 130–131)

406. D. A claim in regard to informed consent is now usually added to charges of malpractice. Informed consent must be based on adequate information concerning therapy, available alternatives, and known and unknown collateral risks. It must be freely given, and not induced in a frenzied situation. (Ref. 3, p. 2111)

407. E. (Ref. 8, p. 670)

408. B. There is no relationship between compliance with diabetic regimens and common demographic characteristics, personality type, and general knowledge of health and illness. ECT leads to an increase in epinephrine, cortisol, and growth hormone, all of which antagonize the effects of lithium. (Ref. 3, p. 1220)

409. E. The diagnostic criteria for a diagnosis of psychological factors affecting physical conditions are important in DSM III. In DSM II, in contrast, psychological factors were considered important only in initiating disorders. Symbolism is not a diagnostic criterion, but rather a psychodynamic mechanism. (Ref. 5, p. 334)

410. B. Obesity is associated with low socioeconomic status in the United States. (Ref. 3, p. 1181)

411. D. Electromyogram is used for the electro diagnosis of peripheral nerve and muscle disease. (Ref. 2, pp. 167–175)

412. C. Thyroid potentiation has been reported in the treatment of breakthrough episodes during lithium prophylaxis during maintenance treatment. (Ref. 2, pp. 924–928)

413. C. Studies of the Mental Hygiene and Child Guidance Movements have not substantiated a reduction in major mental illness in adulthood. Evidence for genetic factors in many disorders makes primary prevention less feasible. (Ref. 3, pp. 2067–2070)

414. B. Cognitive therapy is generally not considered a dynamic psychotherapy. However, increasingly cognitive therapies are becoming integrated with psychodynamic theory and practice. (Ref. 8, pp. 533–538)

415. B. Frugality, punctuality, meticulousness, sadistic rage, problems with control, submission, and defiance are common in people fixated at the anal stage. Because of their use of reaction-formation these characteristics may exist side-by-side with their opposites. In psychoanalytic theory, fixation at any stage of early development is believed to lead to particular personality characteristics. The fixation may occur either because of excessive gratification or because of deprivation, as experienced by the child in terms of his individual needs. (Ref. 6, pp. 14–15)

416. E. The confusion of "b" and "d" is called strephosymbolia and is caused by confusion about position in space. While many children with this disorder display normal intelligence, good math achievement, dyspraxia, spelling and writing difficulties, the clinical picture may vary considerably. (Ref. 3, pp. 1790–1796)

417. B. She meets the criteria for a major depressive episode. She has melancholia: early morning awakening, anorexia, weight loss, and psychomotor retardation. Cancer of the pancreas and retroperitoneal lymphoma are notorious for causing depression. At this point, however, there is no evidence she has cancer. (Ref. 4, pp. 309–314)

418. C. It is not possible to make an axis II diagnosis from the incomplete history given. This does not mean there is not an axis II disorder. It is important to note that DSM III uses five axes. Axis II is for the diagnosis of personality (or, in children, developmental disorder). Axis III is for a traditional medical diagnosis. Axes IV and V, which are coded numerically from 0 to 7, are prognostic, indicating intensity of psychosocial stress—as would be experienced by a "normal" person—and highest level of functioning in the past year. (Ref. 5, pp. 15–21)

419. B. Every depressed patient should be asked about suicide. (Ref. 2, p. 575)

420. E. Suicidal risk increases with depressed mood, especially if vegetative signs are present, as in this case. The risk will become even greater as she improves, and becomes more energetic. Indeed, if she decides to kill herself, her mood may improve remarkably. (Ref. 4, p. 313)

421. E. Somatization disorder (Briquet's syndrome) usually begins in the 20s and runs a fluctuating course, although the patient is rarely free of symptoms. These people do not have symptoms of melancholia. Although they may include conversion phenomena in their list of complaints, the diagnosis of conversion disorder should be reserved for those in whom conversion symptoms are primary. Hypochondriasis is characterized by unrealistic interpretations of physical signs and symptoms as abnormal, and by an obsessive fear that one is ill. (Ref. 5, pp. 261–264)

422. C. Amitryptiline also blocks the reuptake of serotonin. It should be noted that in some groups of severely depressed patients, norepinephrine seems normal, but serotonin (and its breakdown product 5HIAA in the spinal fluid) are low. In other groups, the biochemistry seems normal. (Ref. 8, p. 595)

423. C. When a person who is taking a monoamine oxidase inhibitor eats food containing the amino acid tyramine, he may have a hypertensive crisis. (Ref. 1, p. 515)

424. A. This series of questions involves descriptions of defense mechanisms. By definition, defense mechanisms act automatically, out of our awareness (unconscious) to alleviate anxiety and handle conflict. We all use defense mechanisms. Certain ones, however, especially if overused, will distort reality or produce symptoms. It is most important that the student understand defense mechanisms well. (Ref. 1, pp. 143–144)

425. B. Repression, which is a form of forgetting, should be distinguished from denial, and from suppression. Suppression is a conscious coping mechanism, as when a person consciously decides not to think of something. (Ref. 1, pp. 143–144)

426. D. It is important to distinguish projection from displacement.
427. C. Projection is more primitive and develops during the oral stage of development. It distorts reality and is seen in paranoid conditions. (Ref. 1, pp. 143–144)

428. B. The Durham rule, once regarded as breakthrough in
429. C. forensic psychiatry, has now been abandoned. The
430. A. M'Naghten rule is the traditional rule for determining insanity. It addresses itself to whether the defendant knew the nature and quality of his act and that it was wrong. In many states it is now supplemented by the American Law Institute criteria. Brigg's law holds that a person is not responsible if, because of mental disease, that person is unable to appreciate wrongfulness of his or her conduct. (Ref. 8, pp. 642–643)

431. B. Both barbiturates and benzodiazepines, as do most sleep
432. B. medications, decrease the time of REM sleep. This is
433. A. followed by a return to baseline with continued administration. Finally, if meds are discontinued, REM time will actually show a rebound increase. Reserpine is one of the few drugs that increases REM sleep. (Ref. 4, pp. 159–160)

434. B. Current psychoanalytic understanding suggests that people
435. A. with borderline personalities are fixated at the separation-
436. C. individuation stage of development. They lack evocative
437. A. memory and divide self and object representations into
438. B. "all good" and "all bad." Unlike people who are
439. D. psychotic, their reality testing is intact. People with narcissistic personality disorders, in contrast, lack a coherent sense of self. The schizotypal personality disorder is more common among first-degree biological relatives of persons with schizophrenia. (Ref. 6, pp. 64, 124, 174)

440. B. A fluctuating level of consciousness characterizes delirium. (Ref. 5, p. 103)

441. A. In dementia, there is a gradual loss of intellectual functioning. (Ref. 5, p. 107)

442. D. Hallucinosis is characterized by auditory hallucinations in a state of clear consciousness. (Ref. 5, p. 132)

443. A. On average, both REM and NREM sleep decrease slightly with advancing age. (Ref. 4, p. 156)

444. C. At least half of newborns' sleep is REM sleep. Although scoring of sleep may be problematic in young children, there has been repeated confirmation, and REM sleep may represent a developmentally primitive state. (Ref. 4, p. 156)

445. A. Middle adulthood is an important transition, which for some can be a crisis of mid-life. (Ref. 3, p. 2005)

446. E. Infants can respond periodically to sounds as early as the fourth week. Eye crinkling combined with "reflex" grimacing signals the social smile and the end of the "pre-psychologic" phase. (Ref. 8, p. 59)

447. A. This physician made some of the most common mistakes in giving an antidepressant: (1) the dose is too low; (2) the physician gave a lethal amount (to someone who is depressed and therefore possibly suicidal); and (3) recent studies indicate an additive treatment effect for combining optimal pharmacotherapy with appropriate psychotherapy. (Ref. 1, pp. 301–302)

448. A. Anxiety was due to fear of impending death. Depression resulted from diminished self-esteem from the heart attack. Most management problems originated from denial of illness and inappropriate euphoria, sexual or hostile behavior. (Ref. 8, p. 629)

449. A. (Ref. 1, p. 302)

450. A. The MMPI is a personality profile. All the others are projective tests. (Ref. 2, pp. 475–491)

451. E. All these tests are used to assess brain damage. The Bender-Gestalt is useful when no pathology is suspected. (Ref. 4, pp. 58–64)

452. E. Acute choreoathetotic reactions appear similar to tardive dyskinesia and develop when relatively high doses of potent neuroleptic drugs are rapidly withdrawn. This movement usually only lasts a few days. (Ref. 4, pp. 496–497)

453. D. Milligram for milligram, the piperazines are more potent therapeutically than the aliphatic or piperidine phenothiazines. Sedation and orthostatic hypotension are more common with the aliphatics, and anticholinergic effects are more common with piperidines. (Ref. 7, p. 417)

454. E. Disulfiram (Antabuse) interferes with the metabolism of alcohol. This results in acetaldehyde poisoning. The most serious consequence is severe hypotension. (Ref. 7, p. 475)

455. D. 5-OH tryptamine is serotonin. It is important to distinguish the indolamines from the catecholamines (dopamine, norepinephrine, and epinephrine). Together, they comprise the biogenic amines. (Ref. 2, p. 869)

456. A. Antidepressants seem specific for the panic attacks—as opposed to the anticipatory anxiety—of agoraphobia. There is also indication of efficacy in chronic pain syndrome and bulemia nervosa. Results on anorexia nervosa are more problematic. (Ref. 3, pp. 1636–1637)

457. B. This personality profile is often used as a screening instrument to alert the family practitioner to problem areas he might profitably investigate. (Ref. 8, pp. 224–226)

458. B. The principal diagnosis may be an axis II disorder, such as borderline personality disorder. In such instances, place the phrase (principal diagnosis) after the condition. A mental disorder is considered to exist only when there is distress (symptoms) or disability (decline in level of functioning). Consequently, a disturbance limited to the relationship between the individual and society is not conceptualized as a mental disorder. (Ref. 5, pp. xxii, 19)

459. A. Delirium is less common in the middle-aged than in children and old people. (Ref. 5, pp. 100–104)

460. A. Delusions may occur in many conditions in addition to schizophrenic disorders. Indeed, even to be counted as a diagnostic criterion for schizophrenic disorder, a paranoid delusion must be accompanied by a hallucination. (Ref. 5, p. 395)

461. A. These criteria constitute the basis for challenging the validity of a will. (Ref. 3, p. 2114)

462. E. In its Model Penal Code (1955), the American Law Institute recommended the following test of criminal responsibility: a person is not responsible for criminal conduct if, at the time, as a result of mental disease or defect, he lacked substantial capacity, either to appreciate the criminality of his conduct or to conform his conduct to the requirements of the law. This does not apply to abnormality manifested only by repeated criminal or otherwise antisocial acts. (Ref. 8, p. 643)

463. E. Treatment response is often a failure to achieve adequate screen levels, either through inadequate doses or individual metabolism. Supplementation with T3 or lithium should be attempted before switching to another antidepressant. (Ref. 3, p. 1637)

464. E. A 24-hour urine creatinine should be ordered if there is any reason to be concerned about renal function. Other laboratory tests to be ordered are an EKG and a pregnancy test, if there is any possibility that the patient is pregnant. Lithium has effects on all these areas. (Ref. 1, p. 518)

465. A. The essential feature of this disorder is a pervasive pattern of instability of mood, interpersonal relationships, and self-image beginning in early childhood. (Ref. 5, pp. 346–348)

466. A. Some research indicates that exposure to light for 2 to 4 hours anytime during the day may be effective. Patients should not look directly at the light, but only glance at it from time to time. (Ref. 1, p. 532)

References

1. Kaplan HI, Sadock BJ: *Synopsis of Psychiatry Behavioral Sciences Clinical Psychiatry,* ed 5. Baltimore, Williams & Wilkins Co, 1988.

2. Kaplan HI, Sadock BJ: *Comprehensive Textbook of Psychiatry,* Vol 1, ed 5. Baltimore, Williams & Wilkins Co, 1989.

3. Kaplan HI et al.: *Comprehensive Textbook of Psychiatry,* Vol 2, ed 5. Baltimore, Williams & Wilkins Co, 1989.

4. Nicholi AM Jr: *The New Harvard Guide to Psychiatry.* Cambridge, MA, The Belknap Press of Harvard University Press, 1988.

5. *Diagnostic and Statistical Manual of Mental Disorders,* ed 3. (DSM-III-R) American Psychiatric Association, 1987.

6. Moore BE, Fine BD: *Psychoanalytic Terms and Concepts.* New Haven, The American Psychoanalytic Association and Yale University Press, 1990.

7. Wang RIH: *Practical Drug Therapy.* Milwaukee, Medstream Press, Inc, 1987.

8. Goldman HH: *Review of General Psychiatry,* ed 2. Norwalk, CT, Appleton & Lange, 1988.

5 Surgery

Michael H. Metzler, MD

DIRECTIONS (Questions 467–504): Each of the questions or incomplete statements below is followed by five suggested answers or completions. Select the **one** that is best in each case.

---

467. The fully developed picture of traumatic (hypovolemic) shock is characterized by all of the following **EXCEPT**

　　A. oliguria

　　B. peripheral vasoconstriction

　　C. increased blood viscosity

　　D. mental dullness

　　E. bradycardia

468. Which of the following is most likely to be transmitted by blood transfusion?

　　A. Non-A, non-B hepatitis

　　B. Lymphoblastic leukemia

　　C. Myeloblastic leukemia

　　D. AIDS

　　E. Scurvy

469. The most useful immunosuppressive drug used in organ transplantation is
   A. cyclosporine
   B. pyrimidine analogs
   C. nitrogen mustard
   D. mitomycin
   E. actinomycin-D

470. Severe hemolytic transfusion reactions are most commonly due to
   A. inability to detect major antigens
   B. bacterial contamination of blood
   C. circulatory overload
   D. hypokalemia
   E. patient or blood sample misidentification

471. During the last 40 years, the incidence of cancer of which of the following has decreased?
   A. Breast
   B. Lung
   C. Colon
   D. Stomach
   E. Bone marrow

472. A 12-year-old boy underwent appendectomy for acute appendicitis. Four days later he showed advanced paralytic ileus secondary to generalized peritonitis. The best treatment, in addition to usual supportive measures, is
   A. high doses of Ilopan and Prostigmin
   B. prompt reoperation to place multiple drains and search for local abscesses
   C. prompt reoperation to drain the obstructed intestine by ileostomy
   D. intestinal intubation and continuous suction
   E. fibrinolysin to ablate adhesions

473. The most common serious complication after gastric resection is
   A. thrombophlebitis
   B. duodenal stump blowout
   C. hemorrhage
   D. sepsis
   E. atelectasis

474. Compartmental hypertension in the lower extremities is associated with all of the following **EXCEPT**
   A. tissue pressure greater than 10 to 20 mm of mercury
   B. arterial insufficiency
   C. venous insufficiency
   D. diabetic gangrene
   E. fasciotomy as the treatment of choice

475. Mallory-Weiss syndrome is characterized by
   A. a tear in the esophagus
   B. a tear in the gastric mucosa
   C. a tear in the mucosa of gastroesophageal junction
   D. bleeding from gastric polyp
   E. capillary hemangiomas of the stomach

476. Regarding ruptured appendix, all are true **EXCEPT**
   A. incidence is higher at extremes of age
   B. it is more frequently seen in the poor
   C. it is localized to periappendiceal area in 95% of cases
   D. early antibiotic therapy of appendicitis prevents onset of rupture
   E. appendectomy should be performed in the presence of rupture

477. Regarding carcinoma developing in patients with ulcerative colitis, all are correct **EXCEPT**
   A. incidence increases with the duration of disease
   B. age at onset determines frequency of cancer
   C. neoplasia arises from pseudopolyps
   D. growths are multiple, flat, and infiltrating
   E. prognosis is poor

478. Which of the following indicates a poor prognosis in acute pancreatitis?
   A. Peak of serum amylase level
   B. Glycosuria
   C. High urinary amylase level
   D. A low FiO_2/PaO_2 ratio
   E. Decreased serum calcium level

479. Villous adenoma of the rectum is best managed by
   A. conservative management of repeated sigmoidoscopy
   B. abdominoperineal resection
   C. local resection of lesion
   D. infusion of electrolytes
   E. chemotherapy

480. Which of the following is removed in a prostatectomy?
   A. The urethral mucosa
   B. The adenomatous enlargements
   C. The sphincter
   D. The prostate itself
   E. Verumontanum

481. The most common indication for kidney transplantation is
   A. hydronephrosis
   B. end-stage glomerulo- or pyelonephritis
   C. tuberculosis
   D. Wilms' tumor
   E. stag horn calculus

482. Which of the following urinary calculi is radiolucent?
   A. Calcium oxalate
   B. Uric acid
   C. Cystine stones
   D. Triple phosphate stones
   E. Mixed stones

483. Mixed tumors of the salivary gland
A. are most common in the submaxillary
B. are usually malignant
C. are most common in the parotid gland
D. usually cause facial paralysis
E. are associated with calculi

484. Thyroid carcinoma
A. often produces hyperthyroidism
B. is usually associated with hypothyroidism
C. is usually euthyroid in state
D. metastases produce hormones
E. occurs in toxic nodules

485. Early septic shock is characterized by all of the following EXCEPT
A. hypermetabolism
B. high cardiac output
C. low systemic vascular resistance (SVR)
D. positive blood culture more than 90% of the time
E. warm extremities

486. Two hours after application of the plaster cast for supracondylar fracture, the patient comes back to emergency room with a complaint of severe pain in the hand. Examination reveals swelling of the fingers and cyanosis. The best course would be to
A. observe the patient
B. administer vasodilators
C. administer analgesics
D. cut open the plaster near the fingers
E. cut open the entire plaster cast immediately

487. The most common nerve injury associated with fracture of humerus is
A. radial
B. median
C. ulnar
D. axillary
E. musculocutaneous

488. All of the following are frequently associated with pelvic fractures **EXCEPT**
  A. internal hemorrhage and hypotension
  B. neurologic damage and sacral foramen fracture
  C. posterior hip dislocation and acetabular fracture
  D. urethral injury in a male patient
  E. urethral injury in a female patient

489. Primary treatment of osteosarcoma is
  A. radiotherapy
  B. antimetabolites alone
  C. amputation and chemotherapy
  D. curettage
  E. immunotherapy

490. The treatment of a patient with tension pneumothorax with respiratory distress is
  A. IV fluids
  B. O_2 administration
  C. respiratory stimulants
  D. immediate needle aspiration of air from the pneumothorax
  E. intubation

491. All are true in the case of carcinoma of the lung **EXCEPT**
  A. it is never asymptomatic
  B. it may present as a coin lesion in the periphery of the lung
  C. hemoptysis is common
  D. dry, brassy, nonproductive cough may be the only symptom
  E. complete or partial obstruction of bronchus may predispose to lung infection

492. The most common cause of superior vena caval syndrome is
  A. bronchogenic carcinoma of right upper lobe
  B. trauma
  C. thyroid carcinoma
  D. mediastinal fibrosis
  E. multinodular goiter

493. X-ray of the chest of a patient with tetralogy of Fallot may show all the following features **EXCEPT**
  A. increased vascularity of lung fields
  B. boot-shaped heart
  C. diminished pulsation of pulmonary artery on fluoroscopy
  D. right ventricular enlargement
  E. shadow of the great vessels in superior mediastinum is narrow

494. Dissecting aneurysms of thoracic aorta are most frequently due to
  A. atherosclerosis
  B. syphilis
  C. medial degeneration
  D. trauma
  E. coarctation of aorta

495. Complications of empyema include all **EXCEPT**
  A. bronchopleural fistula
  B. empyema necessitans
  C. pericarditis
  D. osteomyelitis
  E. pneumonitis

496. Primary treatment of postoperative pulmonary embolism consists of
  A. anticoagulants
  B. inferior vena caval ligation
  C. thrombectomy
  D. pulmonary embolectomy
  E. thrombolytic agents

497. All are true in the case of an abdominal aortic aneurysm
      EXCEPT
      A. the majority are asymptomatic and detected during
          routine physical examination
      B. it may produce tender pulsatile abdominal mass
      C. 95% of the aneurysms arise above the level of the renal
          arteries
      D. plain films of the abdomen frequently show calcification
      E. severe back and flank pain indicates rupture or dissection
          of the aneurysm

498. The treatment of arteriovenous fistula consists of
      A. excision of fistula with reestablishment of continuity of
          vessels
      B. ligation of artery distal to the fistula
      C. amputation of the limb
      D. ligation of vein below the level of fistula
      E. medical treatment only

499. All are true in the case of Raynaud's disorder **EXCEPT**
      A. it is common in emotional young females
      B. it is usually the bilateral and symmetrical involvement of
          hands
      C. it is a vasospastic condition
      D. whenever hands are exposed to cold, they become pale,
          followed by cyanosis, and finally rubor
      E. the radial pulses are always absent

500. Acute mastitis occurs frequently during
      A. childhood
      B. puberty
      C. pregnancy
      D. lactation
      E. menopause

501. Metabolic consequences of partial gastrectomy may include all **EXCEPT**
A. megaloblastic anemia
B. iron-deficiency anemia
C. metabolic alkalosis
D. calcium deficiency
E. steatorrhea

502. Combinations of the following tests are routinely used in diagnosis of an intra-abdominal abscess **EXCEPT**
A. vital signs and temperature record
B. ultrasound
C. arteriogram
D. rectal examination
E. CT scan with IV and oral contrast

503. In a patient who has established deep vein thrombosis three days following a cholecystectomy, the treatment of choice is
A. elevation of legs
B. active exercises
C. anticoagulation
D. thrombolytic agents such as streptokinase or urokinase
E. ligation of the common femoral vein on the side involved

504. ABO blood group compatibility is usually required for transplantation of all of the following **EXCEPT** the
A. heart
B. kidney
C. liver
D. cornea
E. pancreas

DIRECTIONS (Questions 505–538): This section consists of clinical situations, each followed by a series of questions. Study each situation and select the **one** best answer to each question following it.

Questions 505–508: A 40-year-old black man was admitted with a two-day history of central abdominal· pain, nausea, vomiting, and constipation. His past history was normal except for appendectomy at age 10. He drinks alcohol occasionally. His temperature was 99.5 °F, blood pressure 110/70, and his abdomen was distended and tender. Plain x-rays of the abdomen are seen in Figure 5.1.

505. The most likely diagnosis is
    A. acute cholecystitis
    B. acute pancreatitis
    C. intestinal obstruction
    D. self-sealing perforation of peptic ulcer
    E. sickle cell crisis

Figure 5.1

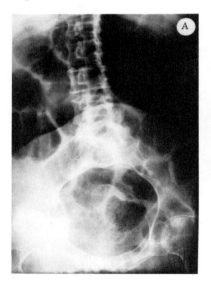
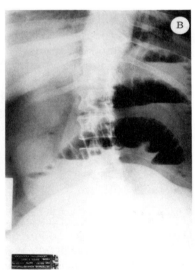

506. All of the laboratory data shown below are consistent with the diagnosis **EXCEPT**
   A. serum sodium 148 mEq/L
   B. total white cell count 10,000/mm³
   C. serum amylase 150 IU%
   D. serum calcium 11 mg%
   E. hematocrit 45%

507. Immediate therapeutic measures include all **EXCEPT**
   A. nasogastric suction
   B. IV fluids
   C. operation
   D. anticholinergic drugs
   E. correction of electrolyte imbalance

508. The procedure of choice is
   A. emergency cholecystectomy
   B. near total pancreatectomy
   C. simple closure of perforated ulcer
   D. exploratory laparotomy and release of adhesions which are probably the cause of mechanical obstruction
   E. no operation; conservative measures will provide relief

Questions 509–512: A 57-year-old man was admitted with a six-month history of intermittent abdominal pain, melena off and on, and a history of altered bowel habit. Barium enema revealed the lesion shown in Figure 5.2.

509. The most likely diagnosis is
   A. stricture due to diverticulitis
   B. granulomatous colitis
   C. carcinoma
   D. lymphogranulomatous stricture
   E. carcinoid tumor

510. A useful diagnostic test that may be used in followup is
- **A.** urinary 5 HIAA
- **B.** serum α-fetoprotein
- **C.** serum CEA levels
- **D.** Frei test
- **E.** serum IgG, IgM levels

Figure 5.2

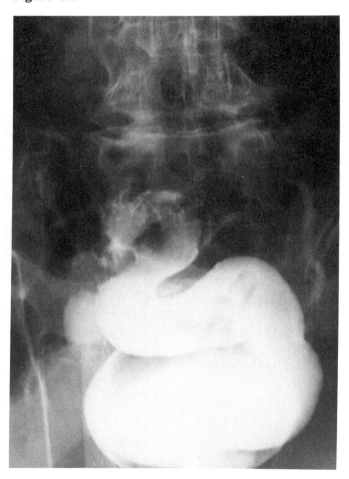

511. The next step in the management is
  A. endoscopic biopsy
  B. visceral angiography to demonstrate source of bleeding
  C. liver scan
  D. inguinal node biopsy
  E. small bowel series to examine for skip lesions

512. Management of this patient includes
  A. salicylazosulfapyridine and steroids
  B. a diet rich in bran and bulk residue
  C. exploratory laparotomy and resection of the lesion
  D. large doses of antibiotics
  E. endoscopic fulguration of the lesion

Questions 513–516: A 25-year-old female presents with a nodule in the left lobe of the thyroid that she discovered accidentally one month prior to her visit. The nodule is not producing any symptoms and is about 2 cm in diameter. In the past she had a brief exposure of radiation, 300 R, to her face for treatment of acne. The I^{131} scan is as shown in Figure 5.3.

Figure 5.3

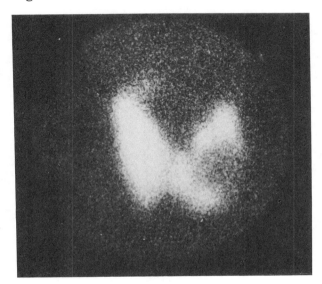

513. Management of this problem requires
   A. observation for any increase in size over a three-month period
   B. suppression therapy with thyroxin
   C. serum T3, T4 levels
   D. serum levels of thyroid antibodies
   E. ultrasound scan

514. Any of the following are reasonable approaches **EXCEPT**
   A. fine needle aspiration cytology of the nodule
   B. left thyroid lobectomy if fine needle cytology is suspicious
   C. open wedge biopsy of the nodule
   D. total thyroidectomy if the biopsy frozen section shows cancer
   E. left thyroid lobectomy only if the frozen section diagnosis is uncertain regarding malignancy

515. If a total thyroidectomy is performed, possible postoperative complications include all **EXCEPT**
   A. hypocalcemia
   B. hypercalcemia
   C. hoarseness
   D. airway obstruction
   E. hemorrhage

516. After surgical therapy the patient should be maintained on
   A. thyroxin
   B. radioactive iodine
   C. methimazole
   D. 5-FU
   E. cis-platinum

Questions 517–520: A 50-year-old obese female underwent a cholecystectomy and T-tube drainage of the common bile duct. On the seventh postoperative day, she developed sudden epigastric and left chest pain. She was short of breath and was sweating profusely. A review of the chart showed that her temperature was between 99 °F and 100 °F for the last two days. EKG is shown in Figure 5.4.

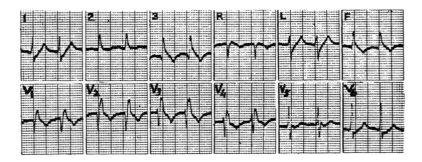

Figure 5.4

517. The most likely diagnosis is
   A. myocardial infarction
   B. pulmonary embolism
   C. gram-negative shock
   D. atelectasis
   E. abdominal wound dehiscence

518. In the management of the patient, which of the following tests is most likely to be useful?
   A. X-ray of the abdomen
   B. Bedside spirometry
   C. Arterial blood gas analysis
   D. Blood cultures X3
   E. Serum LDH and transaminases

519. Which of the following is most useful in corroborating the diagnosis?
   A. Positive radioisotope lung scan
   B. Low arterial pH
   C. Elevated serum LDH
   D. Elevated CBC with left shift in the differential
   E. Hyponatremia

520. The initial management of this patient is
    A. pulmonary embolectomy
    B. urokinase therapy
    C. exploratory laparotomy and drainage of abscess
    D. heparinization
    E. inferior vena caval clipping

Questions 521–524: A 50-year-old man, who was asymptomatic before, presented with a history of acute onset of pain in the left foot and left leg pain of six-hours duration. The left lower extremity was cool up to the mid-calf level and the ankle pulses were absent on the left side. See Figure 5.5.

Figure 5.5

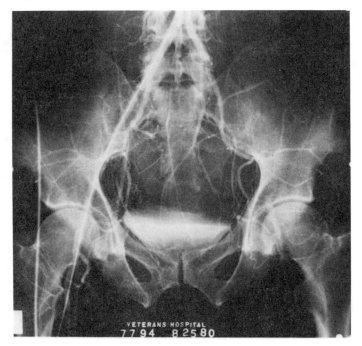

521. The most likely diagnosis is
  A. left iliac embolism
  B. arteriosclerotic occlusion of superficial femoral artery on the left
  C. dissecting aneurysm
  D. saddle embolus
  E. Raynaud's disease

522. In the systemic evaluation of this patient, which of the following findings is most likely to be associated with your previous diagnosis?
  A. A carotid bruit
  B. Atrial fibrillation
  C. Diabetes mellitus
  D. Hypertension
  E. Scleroderma

523. Definitive management of this patient's problem requires
  A. injection of vasodilators
  B. systemic heparinization
  C. embolectomy
  D. hypotensive therapy
  E. sympathectomy

524. The clinical picture described in this patient may be associated with all of the following **EXCEPT**
  A. aortic aneurysm
  B. recent myocardial infarction
  C. secondary hyperthyroidism
  D. diabetic ketoacidosis
  E. initiation of treatment with digitalis and quinidine

Questions 525–529: A 40-year-old woman presents with a lump in her left breast which she noticed on incidental palpation two days prior to the office visit. She has been on birth control pills for six years. The lump is 1.5 cm in diameter and is mobile and painless.

525. Your plan for initial management may include all **EXCEPT**
- **A.** aspiration of fluid if cystic
- **B.** thermography
- **C.** mammography
- **D.** ultrasound
- **E.** aspiration for cytology of solid

Figure 5.6

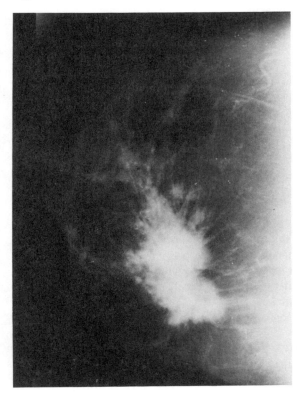

526. Assuming that the mammogram shown in Figure 5.6 reveals patchy calcifications, your next step is
    A. needle biopsy
    B. excisional biopsy
    C. segmental mastectomy
    D. total mastectomy
    E. radical mastectomy

527. If the lesion is malignant, all of the following are factors to weigh regarding adjuvant chemotherapy **EXCEPT**
    A. quadrant of the breast in which the lesion occurred
    B. age of the patient and overall health status
    C. hormonal receptor status of the tumor
    D. presence of tumor in the axillary lymph nodes
    E. a positive bone scan indicating distant metastases

528. Which type of breast cancer has a propensity for bilateral presentation?
    A. Infiltrating ductal carcinoma
    B. Comedo carcinoma
    C. Paget's disease
    D. Lobular carcinoma
    E. Sarcoma of the breast

529. Following mastectomy pathology report of axillary node involvement in 3 of 16 nodes, your plan of treatment is
    A. axillary and mediastinal radiation
    B. oophorectomy
    C. hypophysectomy
    D. chemotherapy
    E. no therapy until recurrence develops

Questions 530–532: A 35-year-old female presented with a history of discharge of blood and mucus from the rectum and tenesmus. Proctosigmoidoscopy revealed a lesion 14 cm from dentate line and a biopsy was performed.

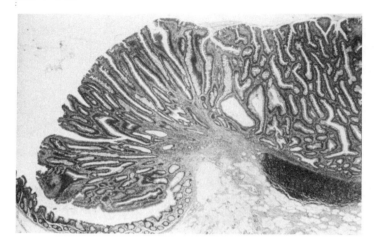

Figure 5.7

530. The histology of the lesion shown in Figure 5.7 is indicative of
  A. juvenile polyp
  B. adenomatous polyp
  C. villous adenoma
  D. pseudopolyp of ulcerative colitis
  E. endometriosis

531. These tumors most often occur in
  A. appendix
  B. cecum
  C. transverse colon
  D. rectum and sigmoid
  E. anal canal

532. If malignancy was noted in the lesion, the treatment of choice is
  A. fulguration
  B. local excision
  C. anterior resection
  D. abdominoperineal resection
  E. colonoscopic excision

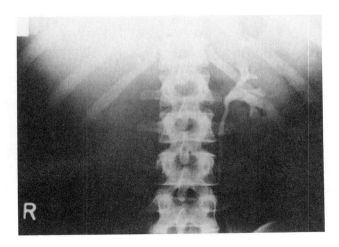

Figure 5.8

Questions 533–535: A 60-year-old male presented with a profuse painless hematuria. At evaluation he was noted to be running a low-grade fever and had lost weight. Clinical examination is unremarkable. An IVP was obtained and is shown in Figure 5.8.

533. The IVP shows
- **A.** hydronephrosis
- **B.** chronic pyelonephritis
- **C.** renal cell carcinoma
- **D.** stone in right kidney
- **E.** acute glomerulonephritis

534. All of the following are useful in further evaluation of the lesion **EXCEPT**
- **A.** ultrasonography
- **B.** arteriography
- **C.** vena cavagraphy
- **D.** CT scan
- **E.** renal scan

535. Assuming that the lesion is a renal cell tumor, treatment is
    A. partial nephrectomy
    B. intracapsular nephrectomy
    C. radical nephrectomy
    D. radiotherapy
    E. chemotherapy

Questions 536–538: A 65-year-old white male with a long-standing history of alcoholism was admitted with deep jaundice (serum bilirubin 8 mg%).

536. In the initial evaluation of this patient, all of the following are acceptable **EXCEPT**
    A. stool guaiac
    B. urine urobilinogen
    C. serum transaminases (SGOT and SGPT)
    D. ultrasonography
    E. intravenous cholangiography

537. Subsequently, a transhepatic cholangiogram (see Figure 5.9) was performed. Based on this study, the most likely cause of jaundice is
    A. alcoholic hepatitis
    B. stones in common bile duct
    C. metastatic cancer probably from stomach or colon
    D. cancer of gallbladder
    E. cancer of head of pancreas

538. If he were to be operated upon, which of the following vitamin deficiencies need immediate correction?
    A. Vitamin K
    B. Vitamin B
    C. Vitamin C
    D. Vitamin D
    E. Vitamin A

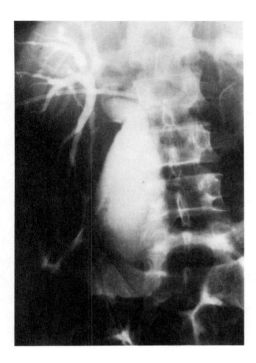

Figure 5.9

DIRECTIONS (Questions 539–546): Each set of lettered headings below is followed by a list of numbered words or phrases. For
each numbered word or phrase select

    A   if the item is associated with **A** only,
    B   if the item is associated with **B** only,
    C   if the item is associated with both **A** and **B**,
    D   if the item is associated with neither **A** nor **B**.

Questions 539–542:
    A. Ulcerative colitis
    B. Granulomatous colitis
    C. Both
    D. Neither

539. High frequency of small bowel involvement

540. Higher risk of cancer

541. Enteric fistulae

542. Proctocolectomy

Questions 543–546:
- **A.** Thyroid
- **B.** Parathyroid
- **C.** Both
- **D.** Neither

543. Produces calcitonin

544. Arises from the fourth branchial arch

545. Arises from the third branchial arch

546. Cancer associated with Sipple's syndrome (MEN IIa)

DIRECTIONS (Questions 547–572): Each group of questions below consists of lettered headings followed by a list of numbered words or statements. For each numbered word or statement, select the **one** lettered heading that is most closely associated with it. Each lettered heading may be selected once, more than once, or not at all.

Questions 547–551:
- **A.** Clindamycin
- **B.** Gentamycin
- **C.** Cephalosporin
- **D.** Bacitracin
- **E.** Chloramphenicol

547. Aplastic anemia

548. Ototoxicity

549. Not absorbed by mouth

550. Antibiotic of choice for prophylaxis of skin organisms in vascular procedures

551. Colitis

Questions 552–555:
    A. Claw hand
    B. Wrist drop
    C. Boutonniere deformity
    D. Mallet finger

552. Injury to ulnar nerve at wrist

553. Injury to extensor tendon in the distal phalanx

554. Injury to central extensor tendon near insertion into middle phalanx

555. Radial nerve injury in upper arm

Questions 556–559: A 52-year-old auto accident victim is brought to the emergency room with the following injuries: open right femur fracture, distended abdomen, severe facial fractures. He is unresponsive to painful stimuli, has gurgling respirations and unequal pupils. His BP is palpated at 60 mmHg; pulse is 160.
    A. Give mannitol 1 g/kg IV
    B. Establish adequate airway
    C. Oxygenate and hyperventilate ($PaCO_2$ 25 to 30 torr)
    D. Give antibiotics and tetanus prophylaxis
    E. Control external hemorrhage and give IV fluids rapidly to treat shock

556. The *first* treatment that should be performed

557. The *second* treatment that should be performed

558. The *third* treatment that should be performed

559. An inappropriate treatment given the patient's present condition

Questions 560–562:
    A. Clark's level I melanoma
    B. A melanoma with a Breslow measurement >4 mm thick
    C. Melanomas of the upper back, arms, neck, scalp (BANS area)

560. Worse prognosis than similar thickness lesion located elsewhere

561. Usually have a 75% or greater five-year survival

562. Usually have occult, metastatic disease at the time of diagnosis

Questions 563–567:
    A. Radical mastectomy
    B. Modified radical mastectomy
    C. Both
    D. Neither

563. Removes pectoralis major muscle

564. Removes nipple-areola complex

565. Removes the long thoracic nerve

566. Removes axillary lymph nodes

567. Never requires adjuvant chemotherapy

Questions 568–572:
    A. Clean (class I operation)
    B. Clean-contaminated (class II operation)
    C. Contaminated (class III operation)
    D. Dirty (class IV operation)

568. The best classification a traumatic laceration can be

569. Usually has <5% infection rate

570. Elective cholecystectomy, for example

571. Ruptured diverticular abscess, for example

572. Consider not closing the skin and subcutaneous tissue

DIRECTIONS **(Questions 573–584):** For each of the questions or incomplete statements below, **one** or **more** of the answers or completions given is correct. Select

    A   if only **1, 2,** and **3** are correct,
    B   if only **1** and **3** are correct,
    C   if only **2** and **4** are correct,
    D   if only **4** is correct,
    E   if **all** are correct.

---

573. Which does(do) not alter the progression of hepatic cirrhosis?
    1. Portocaval shunt
    2. Mesocaval shunt
    3. Sclerosis of esophageal varices
    4. Peritoneo-venous shunt (LeVeen or Denver)

574. Cardinal features of ruptured tubal pregnancy include
    1. amenorrhea
    2. sudden abdominal and pelvic pain
    3. unilateral tender adnexal mass
    4. shock

575. Torsion of the testis should be suspected if
    1. the patient is young
    2. there is sudden pain in the inguinal area
    3. testis is enlarged and tender
    4. hematuria

Directions Summarized				
A	**B**	**C**	**D**	**E**
1,2,3	*1,3*	*2,4*	*4*	*All* are
only	only	only	only	correct

576. The cause(s) of diarrhea in Zollinger-Ellison syndrome is(are)
1. massive gastric acid output into the small bowel
2. acid inactivation of pancreatic juice leading to steatorrhea
3. low intraluminal pH stimulating intestinal peristalsis
4. hypogastrinemia

577. The anatomic division between the right and left lobes of the liver is a plane defined by connection of which structures?
1. Falciform ligament
2. Gallbladder fossa
3. Portal vein
4. Inferior vena cava

578. In a patient with suspected testicular tumor, which of the following is(are) indicated?
1. Tapping the hydrocele and cytologic study of the fluid
2. Measurement of urinary gonadotropins
3. Needle biopsy of the testis
4. Surgical exploration and orchiectomy

579. Features of coarctation of aorta are
1. most often seen just distal to the left subclavian artery
2. difference in BP in upper and lower extremities
3. notching of the ribs
4. left ventricular hypertrophy

580. Indications for coronary bypass operations include
1. severe angina not responding to medical therapy
2. occlusive disease of the left main coronary artery
3. triple vessel disease
4. previous infarction and congestive heart failure

581. Management of venous insufficiency in the leg may include
1. elevation and elastic support stockings
2. ligation and stripping of involved venous system
3. ligation of incompetent perforators
4. ligation of superficial femoral or popliteal veins

582. Which of the following is(are) true in regard to blood-stained discharge from the nipple?
1. It is associated with malignancy in 20% to 30% of cases
2. It may be due to duct papilloma or carcinoma
3. The lesions may be multiple
4. Total mastectomy is the treatment of choice

583. Paget's disease of the nipple
1. is very uncommon and forms 1% of all breast cancers
2. is a primary carcinoma of the ducts
3. is an eczematoid lesion
4. carries a worse prognosis than the average carcinoma of breast

584. Mammography is indicated
1. for annual screening of asymptomatic women ≥50 years old
2. instead of monthly breast self-examination (BSE)
3. for annual follow-up of the remaining breast in women who have undergone mastectomy for cancer
4. instead of annual physical examination of the breasts by a physician

Explanatory Answers

467. E. Bradycardia is not a feature of traumatic hypovolemic shock. Tachycardia is the usual feature. (Refs. 1, p. 43; 2, p. 19; 3, p. 139; 4, p. 139)

468. A. Hepatitis is a serious potential risk (7% to 10%) in patients who have been given blood transfusions. The leukemias are not transmitted by blood transfusion, and scurvy is not a risk. AIDS risk is about 1/250,000 units transfused. (Refs. 1, p. 110; 2, p. 94; 3, p. 133)

469. A. Cyclosporine is the most useful single immunosuppressive drug currently used in transplantation. (Refs. 1, p. 415; 2, p. 232)

470. E. Severe hemolytic transfusion reactions due to major antigen mismatches are usually due to patient identification or bookkeeping errors, rather than inability to adequately type and cross-match blood. Bacterial contamination of properly stored and administered blood is low, as is circulatory overload. Hypokalemia is not a problem. (Refs. 1, p. 110; 2, p. 94; 4, p. 93)

471. D. In the United States during the last 40 years there has been a documented decrease in the incidence of cancer of the stomach. On the other hand, the incidence of lung cancer has increased dramatically during the same period. (Refs. 1, p. 881; 2, p. 423; 3, p. 1177; 4, p. 187)

472. D. The best supportive treatment of paralytic ileus would be intestinal intubation and continuous suction. One should not stimulate the bowel with parasympathomimetic drugs initially. Reoperation should not be planned because it may be possible to control the infection by antibiotics and hope for the ileus to resolve. Fibrinolysins have no place in the management of intestinal ileus. (Refs. 1, p. 913; 2, p. 219)

473. B. Duodenal stump blowout is the most serious and dreaded complication after gastric resection. (Refs. 1, p. 842; 3, p. 492)

474. D. Compartmental hypertension is not seen in association with diabetic gangrene. It is most often seen when there is a delay in arterial repair of longer than six hours' duration. It may be associated with venous obstruction, also; and when the tissue pressure increases above 20 mm of mercury, arterial flow stops. Fasciotomy of both the anterior and posterior compartments should be undertaken, and some surgeons prefer to do a fibulectomy. (Refs. 1, p. 1898; 2, p. 199; 3, p. 2062; 4, p. 149)

475. C. Mallory-Weiss syndrome, which is a cause of upper gastrointestinal bleeding, is usually due to a tear in the mucosa of gastroesophageal junction. It usually follows violent retching and vomiting. (Refs. 1, p. 846; 2, p. 421; 3, p. 1180)

476. D. Antibiotic therapy does not prevent the onset of rupture. Rupture is determined by the site of obstruction and the tension beyond the obstruction. Vascular failure might produce an area of necrosis and consequent perforation. (Refs. 2, p. 464; 3, p. 1318)

477. C. Malignant change in ulcerative colitis does not arise from pseudopolyps. Pseudopolyps are truly either mucosal tags or edematous normal mucosa between ulcerated areas. They do not carry the same premalignant risk as true polyps. (Refs. 1, p. 1015; 2, p. 484; 3, p. 1242)

478. E. A decrease in the serum calcium level is considered to be a poor prognostic sign in acute pancreatitis. Fatty acids precipitate the calcium from the serum as calcium soaps. A *high* FiO_2/PaO_2 ratio indicates respiratory impairment, a poor prognostic sign. (Refs. 1, p. 800; 2, p. 399t; 3, p. 1420; 4, p. 248)

479. C. Villous adenoma of the rectum is best treated by local resection of the lesion. Only if the pathologist finds malignant change is it necessary to proceed with additional treatment. Some of these tumors tend to lose electrolytes and may require infusion, but that is not the primary treatment of the adenoma. (Refs. 1, p. 1003; 2, p. 491; 3, p. 1267; 4, p. 216)

480. B. During prostatectomy, the adenomatous hyperplastic tissue is enucleated from the normal gland. (Refs. 1, p. 1684; 2, p. 842; 3, p. 1774)

481. B. End stage glomerulo- or pyelonephritis leading to chronic renal failure is the most common indication for kidney transplantation. (Refs. 1, p. 408; 2, p. 237; 3, p. 440)

482. B. Pure uric acid stones are radiolucent. Cystine calculi are radiopaque because of their sulphur content. The others contain calcium and are radiopaque. (Refs. 3, p. 1748; 4, p. 233)

483. C. The most common site of mixed tumors of the salivary gland is the parotid. The majority of them are benign and facial paralysis is very uncommon. (Refs. 1, p. 1359; 2, p. 734)

484. C. There is usually no disturbance in the thyroid function in the presence of thyroid carcinoma. Not all metastases produce thyroid hormone. (Refs. 1, p. 602; 3, p. 1634; 4, p. 279)

485. D. A positive blood culture is obtained only about 50% of the time. The physiologic changes are attributed to the release of mediators (biologically active molecules such as cytokines, cachetin, prostaglandins, and complement), not simply the presence of bacteria in the blood. (Ref. 1, p. 20; 4, p. 100)

486. E. The plaster should be cut open entirely and it should include any cotton, wool, or gauze bandages, also. There is a risk of great degree of swelling after reduction and immobilization of the fracture and if this swelling has no place to expand, it could interfere with the blood supply, resulting in ischemia and sometimes even in gangrene of the extremity. Hence, all coverage of the limb should be split immediately and the extremity examined. (Refs. 2, p. 759; 3, p. 2062)

487. A. The radial nerve is most frequently injured in fractures of the shaft of the humerus as the nerve winds around the back of the bone in the spiral groove. (Refs. 1, p. 1443; 2, p. 759; 3, p. 1950)

488. E. Internal hemorrhage and hypotension frequently accompany severe pelvic fracture due to the rich pelvic blood supply and extreme energy necessary to disrupt the pelvis. A sacral foramen fracture often causes injury to the adjacent nerve root. Posterior hip dislocation and acetabular fracture are frequently seen. Male, but not female, urethral injury is often seen. In order to diagnose a partial urethral tear and prevent further injury, a normal digital rectal exam or urethrogram should be performed in a male before bladder catherization. (Refs. 2, p. 762; 3, p. 1974)

489. C. The initial treatment of osteogenic sarcoma is amputation and chemotherapy. (Refs. 1, p. 1510; 2, p. 789; 3, p. 2005)

490. D. In the presence of a tension pneumothorax, treatment should be immediately instituted by inserting a needle into the involved hemithorax. This will immediately equalize the pressure inside the pleural cavity with the atmosphere and correct any mediastinal shift. Insertion of a chest tube is definitive therapy. (Refs. 1, p. 48; 2, p. 181; 3, p. 648; 4, p. 140)

491. A. In a fairly significant percentage of patients, carcinoma of the lung is asymptomatic. It may be discovered accidentally through routine chest x-ray, or may present as metastatic disease elsewhere in the body, or is diagnosed following investigation of some hormonal effect. (Refs. 1, p. 2059; 2, p. 1000; 3, p. 727)

492. A. Bronchogenic carcinoma of the right upper lobe is the most common cause of superior vena caval syndrome. The involved lymph nodes press on the superior vena cava which lies close. (Refs. 1, p. 1728; 2, p. 1000; 3, p. 729; 4, p. 360)

493. A. The lungs of a patient with tetralogy of Fallot will not show an increase in vascularity. In fact, because of the pulmonary stenosis, there may be a decrease in vascularity in the lung fields. (Refs. 1, p. 2224; 2, p. 1090; 3, p. 812)

494. C. Degeneration of the media is the most frequent cause of dissecting aneurysm of the thoracic aorta. These are usually seen in Marfan's syndrome or Erdheim's cystic degeneration. (Refs. 1, p. 1797; 3, p. 981; 4, p. 316)

495. E. Pneumonitis is not a complication of empyema. Empyema may arise in a patient who had pneumonitis. Empyema usually produces collapse of the lung. (Refs. 2, p. 990; 3, p. 676; 4, p. 130)

496. A. The primary treatment of pulmonary embolism consists of anticoagulation. This is usually done with intravenous heparin followed by oral anticoagulation with Coumadin derivatives. Thrombolytic agents are not used in postoperative patients due to fear of bleeding at operative site. (Refs. 1, p. 1740; 2, p. 949; 3, p. 1024)

497. C. Most aneurysms of the abdominal aorta arise below the level of the renal artery. (Refs. 1, p. 1830; 2, p. 401; 3, p. 981; 4, p. 317)

498. A. The ideal treatment for arteriovenous fistula is the excision of the fistula with reestablishment of continuity of both artery and veins. (Refs. 1, p. 1921; 2, p. 1193; 3, p. 942; 4, p. 330)

499. E. The radial pulses are present in Raynaud's disorder. This is a vasospastic disorder affecting the upper limbs and is associated with changes in skin color with cold. (Refs. 1, p. 1926; 2, p. 917; 3, p. 997; 4, p. 330)

500. D. Acute mastitis occurs frequently during lactation. The presence of milk and the trauma associated with breastfeeding are responsible for mastitis. (Ref. 1, p. 538)

501. C. Following partial gastric resection, the patient might show all except metabolic alkalosis. Metabolic alkalosis is usually seen with gastric outlet obstruction and persistent vomiting. It is not an effect of gastrectomy. The megaloblastic anemia is due to failure to produce intrinsic factor, which is necessary for absorption of vitamin B_{12}. Iron-deficiency anemia may be secondary to failure of absorption of iron or to ulceration at the anastomotic site. Calcium deficiency and steatorrhea are often seen. (Refs. 1, p. 956; 2, p. 413; 4, p. 184)

502. C. Intra-abdominal abscess is suggested by spiking temperatures. A rectal exam may diagnose a pelvic abscess by finding a tender mass. Ultrasound and/or CT scan are very useful in locating the abscess and have supplanted arteriography for this purpose. (Refs. 2, p. 204; 3, p. 129; 4, p. 1481)

503. C. The most commonly used method to prevent further thrombus formation is anticoagulation with heparin followed by treatment with Coumadin. The thrombolytic agents are not desirable within the first ten days after operation because of the risk of producing bleeding. All of the other options listed are not recommended. (Refs. 2, p. 66; 3, p. 1017; 4, p. 328)

504. D. Although there have been some reports of improved graft survival with HLA matched corneal transplants, HLA and ABO histocompatibility is not mandatory. (Ref. 1, p. 397; 4, p. 337)

505. C. The abdominal pain, nausea, vomiting, and the x-ray findings showing multiple fluid levels are consistent with intestinal obstruction. In acute cholecystitis and acute pancreatitis, one might see some, but not multiple, fluid-filled loops. There are no visible gallstones. Acute pancreatitis is not likely, because it is not a single loop of dilated proximal jejunum, and there is no history suggestive of it. There is no free air or other evidence of perforation. Sickle cell crisis does not produce multiple fluid levels. (Refs. 1, p. 907; 2, p. 400; 3, p. 1074; 4, p. 199)

506. D. The serum calcium is not elevated in intestinal obstruction. The high levels of serum sodium and hematocrit are probably due to hemoconcentration secondary to vomiting and dehydration. There is usually a slight degree of leukocytosis with intestinal obstruction, but in cholecystitis, pancreatitis, and perforated ulcers these levels are much higher. The serum amylase of 150 units is also consistent with intestinal obstruction, but it is not high enough to diagnose acute pancreatitis. (Refs. 2, p. 441; 3, p. 1077)

507. D. In the acute situations noted in question 505, anticholinergic drugs have no place. Occasionally in acute pancreatitis one might use anticholinergic drugs, but even in a perforated peptic ulcer the immediate treatment does not include antacids and anticholinergic drugs. All others are acceptable measures. (Refs. 2, p. 441; 3, p. 1080; 4, p. 202)

508. D. The obstruction in this individual is predominantly in the small bowel, as seen on the x-ray. The most probable cause is adhesions from the previous appendectomy. Adhesions can produce obstruction immediately after the previous operation or many years later. Exploratory laparotomy is indicated. The other answers are not consistent. In acute cholecystitis, emergency cholecystectomy is only rarely indicated. None of the findings so far confirm the diagnosis of perforated ulcer. Based on the clinical and x-ray findings, conservative measures have no place. It is seldom that mechanical obstruction responds purely to conservative measures. (Refs. 1, p. 909; 2, p. 441; 3, p. 1080; 4, p. 202)

509. C. The most likely diagnosis in a middle-aged individual with that history is a carcinoma. Of all cancers of the large bowel, 70% to 75% occur in the rectosigmoid region. The lesion is too high up for lymphogranuloma stricture, and also rectal strictures due to lymphogranuloma generally are seen in the female. There are no diverticula in the segment nearby, which makes it unlikely that the stricture is due to diverticulitis. The symptoms are of short duration, which makes it unlikely to be granulomatous colitis. Carcinoid tumors of the rectum are so uncommon that they are to be diagnosed by elimination. (Refs. 1, p. 1006; 2, p. 490; 3, p. 1277; 4, p. 217)

510. C. In carcinomas of the colorectum, there is usually an elevation of serum CEA levels. CEA levels may be followed after resection as an indicator of recurrence. Rectal carcinoids do not usually produce hydroxytryptamine; therefore, urinary 5 HIAA levels are not elevated. Serum α-fetoprotein levels are a method of evaluation of primary hepatic tumors. The Frei test is useful for diagnosis of lymphogranuloma, but the diagnosis is not under consideration. Serum IgG/IgM levels do not help in the diagnosis. (Refs. 2, p. 490; 3, p. 1277)

511. A. Further investigation should include endoscopy and a biopsy. The lesion probably cannot be reached with the conventional sigmoidoscope, in which case a colonoscope should be used. Angiography is only useful when the bleeding is brisk, and not when there is a small amount of melena. The value of liver scan in evaluation of cancer of the colon is disputed. In late cases of lymphogranuloma, inguinal node biopsies are of no value. Small bowel series is indicated only if the biopsy reveals granulomatous colitis. (Refs. 1, p. 1006; 2, p. 491; 3, p. 1277; 4, p. 217)

512. C. Whatever the nature of the diagnosis, there is a complete obstruction to the flow of barium; therefore, the patient requires laparotomy and resection of this lesion. Salicylazosulfapyridine and steroids are useful only in the early granulomatous disease, and bulk residue foods are useful, again, prior to formation of a stricture in diverticular disease. Antibiotics have no place in the management of any stricture and, since it is a complete obstruction, endoscopic fulguration is not likely to be useful. (Refs. 1, p. 1007; 3, p. 1279; 4, p. 217)

513. E. An ultrasound study would help in the distinction between solid and cystic nodules, and is indicated next. The isotope scan shows that the nodule is a nonfunctioning area of the gland. The nonfunctioning nodules, which are also referred to as "cold" nodules, are more frequently malignant. The patient's history of previous radiation is also a strong risk factor of malignancy in this thyroid. Since the patient is not grossly dysfunctional, serum levels of T3, T4 are not of great value. Because of the risk of malignancy, observation and suppression therapy are not advisable. (Refs. 1, p. 601; 3, p. 1632; 4, p. 278)

514. C. An open wedge biopsy of a thyroid nodule is not indicated; lobectomy is preferred. (Refs. 1, p. 608; 2, p. 332; 3, p. 1633; 4, p. 280)

515. B. Hemorrhage as well as bilateral recurrent nerve injury are possible complications; either may produce acute airway obstruction. Unilateral recurrent nerve injury would produce hoarseness. Hypocalcemia due to parathyroid removal or damage is possible, but not hypercalcemia. (Refs. 1, p. 1736; 2, p. 336; 3, p. 1640; 4, p. 282)

516. A. Many of the thyroid malignancies are dependent on TSH. Treatment with thyroxin will suppress TSH, and thereby prevent the elevation of TSH levels following thyroidectomy. This helps in the suppression of further growth of the tumor. Radioactive iodine is only indicated in functioning follicular carcinoma of the thyroid. Methamizole has no beneficial effect. 5-FU and cis-platinum have not had any trial in the management of thyroid cancers. (Refs. 1, p. 608; 3, p. 1635; 4, p. 281)

517. B. The EKG reveals an acute right heart failure. Note inversion of T-wave in lead 3 and AVF. In the ventricular leads one can also see development of the right bundle branch block. There is also a prominent S-wave. These changes are suggestive of pulmonary embolism. Gram-negative shock does not develop suddenly. Atelectasis might produce shortness of breath, but not the EKG changes. Abdominal wound dehiscence usually presents with a discharge from the wound. (Refs. 1, p. 1736; 2, p. 951t; 3, p. 1021; 4, p. 329)

518. C. Pulmonary artery obstruction results in decreased perfusion of the lungs, hyperventilation, and a consequent decrease in the alveolar carbon dioxide tension (pCO_2). (Refs. 2, p. 948; 3, p. 1021)

519. A. An isotopic lung scan using macroaggregated albumin followed by ventilation scan using 133 Xenon can provide complementary evidence that the perfusion problem is due to pulmonary emboli. A low arterial pH by itself might be due to a retention of CO_2, or might be a metabolic effect. Elevated CBC with left shift indicates some infection and is really nonspecific. Hyponatremia has no great significance in the diagnosis of pulmonary embolism or any of the conditions mentioned in question 517. (Refs. 1, p. 1739; 2, p. 949; 3, p. 1021; 4, p. 329)

520. D. Initial management of this patient is adequate heparinization. Most patients with pulmonary emboli would resolve on adequate anticoagulation. The initial dose would be given intravenously and could be up to 10,000 units. Pulmonary embolectomy is indicated in only a very small group of patients who have massive embolus and are in a critical state. Vena cava clipping is indicated in patients who continue to throw emboli in spite of adequate anticoagulation or in whom heparinization is not advisable. Treatment with streptokinase or urokinase is not recommended in this postoperative patient. (Refs. 1, p. 1740; 2, p. 949; 3, p. 1024; 4, p. 329)

521. A. The sudden onset of ischemic symptoms in a patient who was previously well strongly points to an embolic episode. Patients with arteriosclerotic occlusion usually have some preexisting symptoms of claudication or chest pain, although very occasionally the first evidence might be a progressive occlusion. Dissecting aneurysm usually starts in the aortic segment and produces chest or abdominal pain. A saddle embolus would give bilateral, lower limb ischemic symptoms. Raynaud's disease is a vasospastic disease affecting mostly the upper limbs. (Refs. 2, p. 898; 3, p. 947; 4, p. 323)

522. B. Atrial fibrillation is a significant finding in patients who have an arterial embolus. It is very likely that the embolus arose in the left atrium and resulted in the acute vascular episode in the left lower limb. A carotid bruit only signifies the presence of coexisting atherosclerotic disease. Hypertension is significant in patients who are suspected to have a dissecting aneurysm. Scleroderma is a feature of Raynaud's disease. (Refs. 1, p. 1908; 2, p. 899; 3, p. 949; 4, p. 324)

523. C. Immediate embolectomy, preferably within 4 to 6 hours, is the definite management in patients suspected to have an acute embolism. Heparinization is necessary, but embolization is the definitive management. If operation is delayed longer, there is the risk of muscle necrosis, myoglobinuria, and kidney failure. Vasodilators have no place in the management of occlusive vascular disease. Hypotensive therapy is useful in dissecting aneurysms. Sympathectomy may be of value in Raynaud's disease. (Refs. 1, p. 1908; 2, p. 901; 3, p. 953; 4, p. 324)

524. D. Arterial emboli can arise from aortic aneurysms, where the clot might be dislodged, following recent myocardial infarction when the thrombus on the ventricular surface of the infarct might be dislodged, and in patients with secondary hyperthyroidism. Secondary hyperthyroidism, also called toxic multinodular goiter, produces frequent involvement of the cardiovascular system, and atrial fibrillation is a common feature of this. Finally, the dislodgment of part of the clot from the atrium results in embolism. It is often preceded by the initiation of treatment of atrial fibrillation with digitalis and/or quinidine. Diabetic ketoacidosis is not a predisposing factor. (Refs. 1, p. 1908; 2, p. 900; 3, p. 950; 4, p. 324)

525. B. Aspiration of a cystic breast mass may be all the treatment necessary if the resulting fluid is not bloody and the cyst does not recur. Ultrasound is better than physical examination for distinguishing cystic from solid lesions. Fine needle aspiration of a solid breast mass for cytologic diagnosis is also a reasonable step. Mammography is helpful in distinguishing malignant from benign breast masses and should also be used in screening this patient for other (nonpalpable) lesions. Thermography is of no practical use. (Refs. 1, p. 536; 2, p. 297; 3, p. 555)

526. B. Excisional biopsy is the ideal management. The mammogram findings of patchy calcifications strongly point to malignancy. In the case of a small lump, there is a chance the needle biopsy might miss the lump or not biopsy the representative area; therefore, excisional biopsy is the right procedure. Without a histologic diagnosis of malignancy, it is not justified to do a segmental, total, or radical mastectomy. (Refs. 1, p. 552; 2, p. 315; 3, p. 563; 4, p. 261)

527. A. All factors except location of the lesion are important to consider when deciding upon adjuvant chemotherapy. (Refs. 2, p. 324; 4, p. 266)

528. D. Lobular carcinoma has the highest incidence of bilaterality. (Ref. 3, p. 565)

529. D. Chemotherapy is indicated in this patient with stage IIB breast cancer. Ablative endocrine therapy as well as "watchful waiting" is not indicated. (Refs. 1, p. 568; 2, p. 280; 3, p. 570; 4, p. 266)

530. C. The histologic lesion shown is a villous adenoma. The appearance such as the leaves of a book is typical of villous lesions. Adenomatous polyps show mucus-secreting cells. (Refs. 1, p. 1004; 3, p. 1265; 4, p. 216)

531. D. Villous tumors are most commonly seen in rectum and sigmoid. Their incidence in other areas is very low. (Refs. 1, p. 1003; 2, p. 489; 3, p. 1264; 4, p. 215)

532. C. Incidence of malignancy in villous adenomas is high and averages about 40%. The operation depends on the size and the location. Anterior resection would be most desirable in this particular patient since the lesion is 14 cm from the dentate line. (Refs. 1, p. 1003; 2, p. 491; 3, p. 1256; 4, p. 217)

533. C. The lesion shown in the IVP is a renal cell carcinoma on the right side. The right renal pelvis is considerably deformed and pushed way up above. Normally, the right renal pelvis is below the left. There is a large, soft tissue mass at the lower pole. There is no evidence of dilated renal pelvis, which one would expect with hydronephrosis or cortical atrophy. (Refs. 1, p. 1650; 2, p. 835; 3; p. 1764)

534. E. Renal scan has no value in evaluation of the lesion. When a renal mass is first diagnosed on intravenous pyelography, ultrasonography is carried out to determine if it is cystic or solid. Arteriography and venography would help in confirming the diagnosis of a solid tumor and to evaluate if it has spread through the renal vein into the vena cava. A CT scan would evaluate metastatic disease or extension. Renal scan, on the other hand, has no value. (Refs. 1, p. 1651; 2, p. 834; 3, p. 1762)

535. C. The treatment of renal cell tumor is radical nephrectomy. Partial nephrectomy would only be considered when there are bilateral tumors or the tumor arises from a solitary kidney. Since it is malignant, one should not do an intracapsular nephrectomy. Radiotherapy and chemotherapy are palliative procedures. (Refs. 1, pp. 1651–1652; 2, p. 834; 3, p. 1755)

536. E. In the evaluation of the jaundiced patient with a serum bilirubin of 8 mg%, intravenous cholangiography has no value. This particular technique of imaging biliary tract is not applicable when the serum bilirubin is above 3 or 3.5 mg. All others are useful. A stool guaiac may suggest the presence of a malignancy either of the ampulla or elsewhere in the GI tract. The absence of urobilinogen in urine indicates a complete obstruction to the bile ducts. Transaminases help in the exclusion of hepatic cellular disease and ultrasonography is helpful in defining dilated ducts or in large masses. (Refs. 2, p. 602; 3, p. 1430)

537. E. The transhepatic cholangiogram shows a distended gallbladder and dilated common and intrahepatic ducts, which is a typical feature of carcinoma of the head of the pancreas. Stones in the common duct are usually associated with stones in the gallbladder, in which case the gallbladder would be shrunken. Metastatic cancer producing obstructive jaundice usually is at the porta hepatis or due to intrahepatic nodules. The presence of a distended gallbladder would rule out cancer of the gallbladder. (Refs. 1, p. 1195; 2, p. 601; 3, p. 1431; 4, p. 248)

538. A. In the patient with obstructive jaundice, when the bile does not reach the intestinal tract, absorption of fat is impaired. This results in impaired absorption of fat-soluble vitamins such as A, D, E, and K. Vitamin K deficiency would result in prolonged prothrombin time; this would increase hemorrhage during the operative procedure. Hence, vitamin K should be administered to jaundiced patients due to undergo operative procedures so that the prothrombin time is normal. (Refs. 1, p. 1194; 3, p. 1430)

539. B. Granulomatous colitis has been associated with a high incidence of small bowel involvement. On the other hand, ulcerative colitis is usually confined to the colon and rectum and any involvement is only limited to the terminal ileum; the so-called back wash ileitis. (Refs. 1, p. 921; 2, p. 483t; 3, p. 1246; 4, p. 222)

540. A. There is a higher risk of cancer developing in patients with ulcerative colitis of long-standing duration. (Refs. 1, p. 1015; 2, p. 483t; 3, p. 1246; 4, p. 222)

541. B. Enteric fistulae are a feature of granulomatous colitis and are similar in etiology to those due to Crohn's disease of the small bowel. (Refs. 1, p. 922; 2, p. 483t; 3, p. 1246; 4, p. 222)

542. A. Proctocolectomy is a curative procedure for treatment of ulcerative colitis. On the other hand, treatment of granulomatous colitis requires segmental resection and there is no permanent cure. (Refs. 1, p. 1118; 2, p. 483t; 3, p. 1246; 4, p. 222)

543. A. The interfollicular cells of the thyroid produce calcitonin. Although this hormone is active in calcium homeostasis, it arises from the thyroid, and not the parathyroid. (Refs. 1, p. 337; 3, p. 1636; 4, p. 277)

544. C. Both the thyroid and parathyroid gland derive from the fourth branchial arch. However, only the superior pair of parathyroids are from the fourth branchial arch. (Refs. 1, p. 620; 3, p. 1614; 4, p. 276)

545. B. The lower pair of the parathyroid glands arise from the third branchial arch, along with the thymus. (Refs. 1, p. 620; 3, p. 1644; 4, p. 276)

546. A. The association of medullary thyroid cancer with pheochromocytoma is the original description of Sipple's syndrome. Occasionally, hyperparathyroidism is associated with Sipple's syndrome; the lesion in the parathyroid is not a cancer, but an adenoma. (Refs. 1, p. 612; 2, p. 360; 3, p. 1636)

547. E. The use of chloramphenicol has been associated with development of aplastic anemia in sufficient numbers that this risk should be kept in mind. (Refs. 1, p. 443; 3, p. 194)

548. B. Gentamycin, like other aminoglycosides, can produce ototoxicity. (Ref. 1, p. 271t)

549. D. Bacitracin is not absorbed by mouth. (Ref. 3, p. 195)

550. C. A first-generation cephalosporin is the drug of choice for skin prophylaxis in vascular procedures. (Refs. 1, p. 2301; 3, p. 192)

551. A. Clindamycin-associated pseudomembranous colitis has been reported often enough that this is a serious complication to be kept in mind when this antibiotic is used. (Refs. 1, p. 271t; 3, p. 1254)

552. A. Injury to the ulnar nerve at the wrist results in a claw hand due to the paralysis of the intrinsic muscles of the hand. (Ref. 3, p. 2072)

553. D. Mallet finger is produced by avulsion with or without a piece of bone of the terminal insertion of the extensor tendon. This results in the typical deformity. (Refs. 2, p. 746; 3, p. 1960)

554. C. Injuries to the extensor tendon of the fingers near the insertion into the middle phalanx results in the Boutonniere deformity. (Refs. 2, p. 746; 3, p. 2053)

555. B. A radial nerve injury in the upper arm, such as that due to a fracture shaft of the humerus, results in paralysis of the extensor muscles of the wrist; therefore, wrist drop is seen. (Ref. 3, p. 1950)

556. B. The priorities for treatments of *all* multiple-trauma patients
557. C. are: airway, breathing, control of hemorrhage, and resusci-
558. E. tation from shock. These may be performed simultane-
559. A. ously if adequate personnel are available. While mannitol may be indicated in the treatment of a closed head injury with pending herniation, it is contraindicated while the patient is in shock. (Refs. 1, p. 295; 4, p. 134)

560. C. Thin melanomas (Clark's level I or Breslow <0.76 mm
561. A. thickness) have best prognosis. Thick lesions (Clark's level
562. B. V or Breslow >4 mm) have poor prognosis. Lesions in the BANS area have a worse prognosis than lesions of similar thickness located elsewhere on the body surface. (Ref. 1, p. 516; 3, p. 546)

563. A. Removal of the major pectoralis muscle distinguishes the
564. C. radical mastectomy. Both procedures remove the nipple
565. D. and axillary nodes; neither routinely removes the long
566. C. thoracic nerve (this would produce a winged scapula due to
567. D. denervation of the serratus anterior muscles). Whether or not adjuvant therapy is required depends upon the stage of the disease, not the operation performed. (Refs. 1, p. 555; 4, p. 265)

568. C. A class III is the best a traumatic wound can be. Class I
569. A. cases should have fewer than 5% wound infections. An
570. B. elective cholecystectomy is usually a class II case. A
571. D. ruptured diverticular abscess is a class IV case and
572. D. nonclosure of the skin and subcutaneous fat is a reasonable strategy to decrease the risk of wound infection. (Refs. 2, p. 148; 4, p. 122t)

573. E. Portocaval and mesocaval shunts lessen the chance for fatal hemorrhage from esophageal varices by relieving portal hypertension. Sclerosis treats the varices directly. Peritoneo-venous shunts relieve ascites. All operations have no effect on the progression of liver disease and essentially treat only the symptoms of cirrhosis and portal hypertension. (Refs. 2, p. 533; 4, p. 19)

574. E. All of the features listed are usually seen in ruptured tubal pregnancy. (Refs. 1, p. 807; 3, p. 1802)

575. A. Torsion of the testis is usually seen in young individuals. The patient has a sudden pain in the inguinal region and on examination, the testis is enlarged and is extremely tender. Hematuria is not seen. (Refs. 1, p. 1678; 2, p. 404; 3, p. 1735)

576. A. Diarrhea in Zollinger-Ellison syndrome is due to the first three factors outlined. In Zollinger-Ellison syndrome, there is no hypogastrinemia. (Refs. 2, p. 417; 3, p. 1170; 4, p. 182)

577. C. The right and left lobes of the liver are separated by a plane connecting the gallbladder fossa with the inferior vena cava. (Refs. 1, p. 1054; 2, p. 513; 3, p. 1350)

578. C. With a suspected testicular tumor, tapping of the hydrocele or needle biopsy of the testis should not be done because of the danger of dissemination of the tumor. (Refs. 1, p. 1677; 2, p. 851; 3, p. 1769)

579. E. All are features of coarctation of aorta. (Refs. 1, pp. 2178–2181; 2, p. 1065; 3, p. 789)

580. A. Indications for coronary artery bypass grafting (CABG) include severe angina not responding to medical therapy, left main artery disease, and triple vessel disease. Patients with congestive heart failure and pulmonary hypertension do not respond to CABG. (Ref. 3, p. 887)

581. A. Principles of management of venous insufficiency in the leg include elevation at rest and elastic support while walking, ligation and stripping of any varicose veins involved, and ligation of incompetent perforator veins. Ligation of superficial femoral or popliteal veins is not justified. (Refs. 1, p. 1719; 2, p. 939; 3, p. 1028; 4, p. 328)

582. A. Blood in the discharge from the nipple is associated with malignancy in 20% to 30% of cases. It may be part of chronic cystic mastitis. When associated with tumor, it may be due to a papilloma or a carcinoma, and these lesions may be multiple. Mastectomy for bloody discharge alone is not justified. (Refs. 1, p. 535; 2, p. 302; 3, p. 562; 4, p. 261)

583. A. Paget's disease of the nipple is an uncommon carcinoma of the breast and produces an eczematous ulcerated lesion of the areola. The underlying lesion is a carcinoma of the ducts. (Refs. 1, p. 546; 2, p. 311; 3, p. 563)

584. B. Mammography is useful for annual screening of asymptomatic women ≥ 50 years old and those who have undergone mastectomy. It should be used in addition to monthly BSE and annual physical exam by a physician. (Refs. 2, p. 306; 3, p. 555; 4, p. 268)

References

1. Sabiston DC (ed): *Textbook of Surgery,* ed 13. Philadelphia, WB Saunders Co, 1986.

2. Sabiston DC (ed): *Sabiston's Essentials of Surgery.* Philadelphia, WB Saunders Co, 1988.

3. Schwartz SI (ed): *Principles of Surgery,* ed 5. New York, McGraw-Hill Book Co, 1979.

4. Lawrence PF (ed): *Essentials of Surgery.* Baltimore, Williams & Wilkins, 1988.

6 Obstetrics and Gynecology

Sherwood C. Lynn, Jr, MD, and
Theodore P. Haddox, Jr, MD

DIRECTIONS (Questions 585–611): Each of the questions or incomplete statements below is followed by five suggested answers or completions. Select the **one** that is best in each case.

---

585. Neonatal mortality is
   A. synonymous with stillbirth rate
   B. the number of neonatal deaths per 1000 births
   C. the number of neonatal deaths per 100 live births
   D. the number of neonatal deaths per 1000 population
   E. the number of neonatal deaths in the first seven days of life

586. K-cells in the corpus luteum are
   A. found in cumulus oophorus
   B. found in theca externa
   C. primordial follicles
   D. cells containing alkaline phosphatase
   E. cells containing placental lactogen

587. The following statements are true **EXCEPT** the syncitium (syncytiotrophoblast)
    A. is derived from cytotrophoblast
    B. is a mitotic end stage
    C. contains well-developed Golgi complexes
    D. has abundant endoplasmic reticulum
    E. is composed of Hofbauer cells

588. Human placental lactogen is known to be involved with all but which one of the following?
    A. Lipolysis
    B. Successful pregnancy
    C. Inhibition of maternal gluconeogenesis
    D. Augmentation of maternal glucose uptake
    E. Production by a hepatoma

589. Approximately 50% of the glycerophospholipids (surfactant) is
    A. dipalmitoylphosphatidylcholine
    B. phosphatidylglycerol
    C. phosphatidylinositol
    D. phosphatidylethanolamine
    E. phosphatidate phosphohydrolase

590. During normal pregnancy all but which of the following is true?
    A. The resting pulse rate increases by about 10 to 15 beats/min
    B. The heart is displaced upwards and to the left
    C. Arterial blood pressure and vascular resistance increase
    D. The electrocardiogram shows deviation to the left
    E. Cardiac volume increases by 10%

591. Engagement of the fetal head is
    A. synonymous with fixation
    B. when the occiput has passed through the pelvic inlet
    C. when the widest diameter of the presenting part has passed through the pelvic inlet
    D. only present after labor begins
    E. normally present at 36 weeks gestation in multiparae

592. The level of α-fetoprotein in amniotic fluid and/or maternal serum may be raised in all of the following conditions **EXCEPT**
  A. congenital nephrosis or fetal bladder neck obstruction
  B. open neural tube defects
  C. hydrocephaly
  D. Turner's syndrome (45XO)
  E. esophageal and duodenal atresia

593. The latent phase of labor
  A. follows the phase of maximum slope and precedes the second stage
  B. follows the acceleration phase and precedes the phase of maximum slope
  C. follows the active phase
  D. precedes the active phase
  E. follows the deceleration phase

594. The histological features of a hydatidiform mole include all of the following **EXCEPT**
  A. hydropic degeneration of the villous stroma
  B. absent fetal vessels
  C. aplasia of syncytium
  D. absence of amnion
  E. hyperplasia of cytotrophoblast

595. Concerning amniotic fluid, all but which one of the following is true?
  A. Acute polyhydramnios is commonly associated with monozygous twins
  B. Acute polyhydramnios is more common than chronic polyhydramnios
  C. Polyhydramnios is associated with maternal diabetes
  D. Polyhydramnios is associated with hydropic erythroblastosis
  E. Polyhydramnios is associated with increased perinatal mortality

596. Preeclampsia is defined as the
  A. development of hypertension after 20 weeks gestation
  B. appearance of proteinuria with edema during pregnancy
  C. development of hypertension with albuminuria or edema or both during pregnancy
  D. development of papilledema during pregnancy
  E. occurrence of convulsions not caused by neurological disease during pregnancy

597. Concerning acute pyelonephritis complicating pregnancy and the puerperium, all of the following statements are true **EXCEPT**
  A. it affects approximately 2% of patients
  B. when unilateral, it is most often right-sided
  C. symptoms include anorexia, nausea, and vomiting
  D. *E. coli* is the predominant causative microorganism
  E. there is a concomitant deficiency of autoimmunity

598. Concerning iron deficiency anemia in pregnancy, which of the following statements is true?
  A. One pregnancy can be an important cause of anemia in the next pregnancy
  B. Overt anemia can be adequately treated with 200 mg ferrous gluconate daily
  C. A daily supplement of 30 mg of elemental iron is sufficient to meet the total requirements during pregnancy
  D. With absent iron stores and iron deficiency, anemia can be treated with 60 mg of elemental iron daily
  E. A, B, and C are true

599. Maternal rubella infection during pregnancy may produce all but which one of the following stigmata in the neonate?
   A. Cataracts, glaucoma, and/or microphthalmia
   B. Severe defects of long bone development with failure of closure of the epiphyses
   C. Congenital heart disease, including patent ductus arteriosus and pulmonary stenosis
   D. Meningoencephalitis
   E. Hepatosplenomegaly and jaundice, thrombocytopenia, and anemia

600. The most common organism isolated from puerperal mastitis (parenchymatous inflammation) of the mammary glands is
   A. *E. coli*
   B. *S. aureus*
   C. *Streptococcus fecalis*
   D. *Staphylococcus epidermidis*
   E. *Bacteroides* species

601. Which of the following statements concerning infections with group B, β-hemolytic streptococcal infections is **NOT** true?
   A. Colonization of the female genital tract in late pregnancy is found in about 40% of patients
   B. Early onset neonatal disease is usually associated with premature spontaneous rupture of membranes
   C. Early onset neonatal disease is usually septicemia
   D. Late onset neonatal disease is usually meningitis
   E. Iatrogenic introduction following obstetric manipulation is the main cause of maternal colonization

602. The most common chromosomal abnormality reliably diagnosed by amniocentesis is
A. trisomy 18
B. 13-15/21 translocation
C. trisomy 13
D. 21/22 translocation
E. trisomy 21

603. Which of the following statements is **NOT** true concerning anorexia nervosa?
A. It is a disease of adolescence characterized by severe malnutrition without associated lethargy
B. The most frequent age distribution is between 11 and 21 years
C. 90% of the patients are female
D. There is a pathological obsession with body size resulting in a refusal, or inability, to recognize hunger
E. FSH (follicle stimulating hormone) levels are pathognomonic

604. Cytogenetic studies of hydatidiform moles have indicated that the most common chromosomal composition is
A. 46XY
B. 46XX
C. 45X
D. 47XXY
E. 47XYY

605. A dermoid cyst (benign cystic teratoma) is comprised of
A. squamous epithelium only
B. sweat glands, apocrine glands, and epithelial pearls
C. squamous epithelium with sebaceous glands only
D. all three germ cell layers
E. ectoderm and endoderm only

606. Müllerian duct regression factor is produced by
   A. Leydig cells
   B. Hofbauer cells
   C. K-cells
   D. Sertoli cells
   E. Langhans cells

607. All of the following features are noted in women with testicular feminization **EXCEPT**
   A. mammary aplasia
   B. female phenotype
   C. no Wolffian duct structures
   D. short vagina
   E. no Müllerian duct structures

608. Renal agenesis is associated with all of the following **EXCEPT**
   A. oligohydramnios
   B. low-set ears
   C. small hand
   D. flattened nose
   E. prominent epicanthal folds

609. Ablation of beat-to-beat variation in fetal heart rate may be observed in association with
   A. normal fetal sleep
   B. phenothiazines
   C. conduction anesthesia
   D. atropine
   E. all of the above

610. Clinically, the most sensitive marker for increased androgen production is
   A. acne and increased oiliness of the skin
   B. increased libido
   C. hirsutism
   D. menstrual irregularity
   E. clitoromegaly

611. The following statements in the evaluation of hirsutism are true **EXCEPT**
   A. occasionally testosterone levels are greater than 200 mg/dL in anovulatory patients
   B. enlarged ovaries are not necessary for anovulation and increased androgen production
   C. laparoscopy and ovarian biopsy are indicated procedures in the evaluation of hirsutism
   D. suppression of increased androgens by progestin treatment does not rule out the presence of an ovarian tumor, since ovarian tumors are gonadotropin-dependent and -responsive
   E. the failure of progestin treatment to suppress hair growth and testosterone levels after 6 to 12 months raises the suspicion of adrenal disease or a small ovarian tumor

---

DIRECTIONS (Questions 612–700): For each of the questions or incomplete statements below, **one** or **more** of the answers or completions given is correct. Select
   A   if only **1, 2,** and **3** are correct,
   B   if only **1** and **3** are correct,
   C   if only **2** and **4** are correct,
   D   if only **4** is correct,
   E   if **all** are correct.

---

612. A sinusoidal fetal heart rate pattern is associated with
   1. fetal anemia
   2. Rh-isoimmunized fetus
   3. meperidine
   4. naloxone

613. The human ovary is capable of producing
   1. dehydroisoandrosterone
   2. androstenedione
   3. testosterone
   4. dehydroepiandrostenedione

614. In the male embryo,
1. the primordial germ cell enters the developing gonad during the fifth week
2. the germ cells locate in the medulla, rather than the cortex
3. the germ cells are incorporated into primitive sex cords derived from surface epithelium
4. at birth the sex cords are solid

615. Placental transfer may be affected by the
1. concentration gradient
2. maternal and/or fetal blood flow
3. exchange area available
4. amount of the substance metabolized during transfer

616. The change from fetal to neonatal circulation involves
1. constriction of the umbilical vessels, the foramen ovale, and the ductus venosus after birth
2. a fall in systemic blood pressure
3. expansion of the fetal lungs
4. reversal of blood flow in the ductus arteriosus

617. Which of the following statements concerning the determination of fetal sex is true?
1. In the absence of the testis, female differentiation ensues irrespective of genetic sex
2. Genetic sex is established at the time of fertilization
3. Gonadal sex is determined by the action of a locus on the Y chromosome
4. The development of ambiguous sex is the result of abnormal androgen representation in utero

618. Common errors in the resuscitation of the newborn include
1. failure to check equipment
2. use of a cold resuscitation table
3. unsuccessful intubation
4. inadequate ventilation

Directions Summarized				
A	**B**	**C**	**D**	**E**
1,2,3	*1,3*	*2,4*	*4*	*All* are
only	only	only	only	correct

619. Which of the following may be associated with placental abruption?
1. Renal failure
2. Respiratory failure
3. Disseminated intravascular coagulation defects
4. Heart failure

620. Which of the following statements is(are) correct concerning the pathology of cervical carcinoma?
1. Squamous cell carcinoma is as common as adenocarcinoma
2. The majority of lesions develop at the squamocolumnar junction
3. Adenosquamous lesions carry the best prognosis
4. Squamous cell carcinoma accounts for the majority of cases

621. Included in the differential diagnosis of acute pelvic inflammatory disease limited to the pelvis is(are)
1. acute appendicitis
2. acute urinary tract infection
3. adnexal torsion
4. lower lobe pneumonia

622. Factors associated with intrauterine growth retardation include
1. fetal intrauterine infections
2. fetal chromosomal abnormalities
3. placental chorioangiomata
4. maternal age

623. Asymptomatic bacteriuria in the pregnant woman is
1. present in 15% to 17% of all pregnant women
2. arbitrarily defined as the presence of 100,000 bacteria/mL of urine
3. predominantly caused by *S. fecalis*
4. associated with the development of 30% of cases with symptomatic urinary tract infections

624. Concerning ovulation, which of the following statements is (are) true?
1. Available information suggests that the rise in estradiol in the late follicular phase sets off the LH surge
2. Administration of exogenous estrogens does not affect gonadotropin release
3. Abnormal levels of androgens may suppress the cyclic center and the outpouring of GnRH
4. The midcycle rise in FSH is independent of the physiological effect of the LH surge

625. The characteristics of Turner's syndrome (karyotype XO) are
1. short stature
2. a female child with bilateral inguinal hernia
3. hypergonadotropic hypoestrogenic amenorrhea
4. elevated gonadotropins with the presence of ovarian follicles and amenorrhea

626. Pruritus vulvae may be associated with which of the following?
1. Vaginal discharge
2. Systemic diseases
3. Psychosomatic illness
4. Chronic vulval dystrophies

627. Before a sperm can fertilize an oocyte, it must undergo
1. a meiotic division
2. capacitation
3. second meiotic metaphase
4. acrosome reaction

246 / Clinical Sciences

Directions Summarized				
A	B	C	D	E
1,2,3	*1,3*	*2,4*	*4*	*All* are
only	only	only	only	correct

628. Although the yolk sac is a vestigial structure, its embryonic development is essential as
    1. its blood vessels become the umbilical vein and arteries
    2. the dorsal part of the yolk sac is incorporated into the embryo as an endodermal tube
    3. it forms the urachus
    4. the earliest evidence of embryonic hematopoiesis appears in the wall of the yolk sac in the third week

629. The round ligament
    1. can be considered as of Wolffian duct origin
    2. terminates in upper part of labium majus
    3. supports the uterus
    4. corresponds embryologically to the gubernaculum testis

630. Branches from the femoral artery that participate in the pelvic collateral circulation include the
    1. medial femoral circumflex artery
    2. inferior gluteal artery
    3. lateral femoral circumflex artery
    4. external pudendal artery

631. It is important for the surgeon to know the relations of the hypogastric artery. Immediately posterior to the hypogastric artery are the
    1. ureter
    2. hypogastric vein
    3. external iliac vein
    4. origin of the common iliac vein

632. The cardinal ligaments are
1. condensations of endopelvic fascia
2. below and continuous with the loose cellular tissue that invests the ureter
3. known to run from the pelvic wall in the neighborhood of the sacroiliac joints to the sides of the cervix and lateral aspects of the vagina
4. below and continuous with the loose cellular tissue that invests the uterine vessels

633. The gonads are derived from the
1. coelomic epithelium
2. allantois
3. medial aspect of the urogenital ridge
4. neuroectoderm

634. The causes of delayed puberty include
1. gonadal dysgenesis
2. regional ileitis
3. Kallmann's syndrome
4. sickle cell disease

635. The catecholamine neurotransmitters that regulate hypothalamic function include
1. dopamine
2. tryptophan
3. norepinephrine
4. serotonin

636. In humans, prolactin secretion is stimulated by
1. serotonin antagonists
2. insulin-induced hypoglycemia
3. L-dopa
4. thyrotropin-releasing hormone

Directions Summarized

A	B	C	D	E
1,2,3	*1,3*	*2,4*	*4*	*All* are
only	only	only	only	correct

637. Indications for hormone replacement in women at the menopause include
    1. prevention of atherosclerotic heart disease
    2. relief of vasomotor symptoms
    3. prevention of osteoporosis
    4. reversal of atrophy of urethral mucosa

638. Contraindications to the prescribing of oral contraceptives include
    1. acute cholestatic liver disease
    2. familial hyperlipidemia
    3. undiagnosed abnormal uterine bleeding
    4. a past history of thrombophlebitis

639. During the luteal phase of the normal menstrual cycle,
    1. FSH levels are lower than in the follicular phase
    2. androstenedione levels are higher than at midcycle
    3. progesterone levels are maximum at the middle of the luteal phase
    4. testosterone levels are higher than at midcycle

640. The principal androgens involved in women with evidence of hyperandrogenism are
    1. testosterone
    2. androsterone
    3. androstenedione
    4. etiocholanolone

641. Turner's syndrome is associated with short stature and sexual infantilism. Other frequent features include
    1. micrognathia
    2. high arched palate
    3. epicanthal folds
    4. hypertrichosis

642. The production of estrogens by the placenta appears to depend upon
  1. the availability of androgen precursors from the fetus
  2. progesterone in the maternal circulation
  3. the availability of androgen precursors from the mother
  4. progesterone in the fetal circulation

643. Drugs that affect the metabolism of steroidal antifertility agents include
  1. diazepam
  2. rifampicin
  3. phenytoin
  4. ampicillin

644. Precocious puberty is associated with
  1. puberty changes before the age of eight years
  2. Von Recklinghausen's disease
  3. Albright's syndrome .
  4. Noonan's syndrome

645. Side effects associated with clomiphene citrate therapy include
  1. nausea
  2. vasomotor disturbances
  3. abdominal discomfort
  4. visual symptoms

646. Hyperprolactinemia can be associated with
  1. cervical spinal lesions
  2. stress
  3. hypothyroidism
  4. anesthesia

647. A congenital enzyme block due to 11 β-hydroxylase deficiency is associated with
  1. excessive corticosterone production
  2. virilization
  3. salt depletion
  4. excessive deoxycorticosterone production

Directions Summarized				
A	**B**	**C**	**D**	**E**
1,2,3	*1,3*	*2,4*	*4*	*All* are
only	only	only	only	correct

648. In the American College of Obstetrics and Gynecology adaptation of the Priscilla White classification of diabetes in pregnancy, class C has which of the following characteristics?
   1. Vascular disease is present as evidenced by retinitis
   2. Onset of diabetes occurs between the ages of 10 and 20 years
   3. Nephritis is present
   4. Duration of the disease is 10 to 19 years

649. Lumbar epidural analgesia
   1. produces pain relief in the first stage of labor by blocking sensory fibers from T10-L2
   2. may produce maternal hypotension by blocking sympathetic nerve fibers
   3. is contraindicated in patients with Von Willebrand disease
   4. may be associated with convulsions, total spinal anesthesia, and fetal bradycardia

650. Fetal intrauterine growth retardation is associated with
   1. teenage pregnancy
   2. smoking
   3. maternal hypertension
   4. Down syndrome

651. Which of the following factors predispose an infant to kernicterus?
   1. Hypothermia
   2. Sepsis
   3. Hypoglycemia
   4. Vitamin K analogs

652. Amniocentesis is indicated for the diagnosis of which of the following autosomal inherited disorders?
  1. Hurler's syndrome
  2. Tay-Sachs disease
  3. Pompe's disease
  4. Lesch-Nyhan syndrome

653. In pregnancy, the oral glucose tolerance test
  1. is not influenced by the duration of pregnancy
  2. is altered by a slight elevation of blood sugar in the postabsorptive state
  3. always furnishes identical results when repeated in the same patient
  4. is altered by a lag in the return to normal

654. Glucose-6-phosphate dehydrogenase deficiency is
  1. a hereditary metabolic disorder carried on the X chromosome
  2. associated with a high incidence of urinary tract infections in pregnancy
  3. a cause of fetal hydrops
  4. a cause of neonatal jaundice

655. Ritodrine, which is used to arrest labor, is
  1. a β-mimetic agent
  2. an analog of isoxsuprine
  3. never given in an intravenous dosage in excess of 400 μg/min
  4. not a cause of maternal tachycardia

656. Factors other than the serum bilirubin concentration that contribute to the development of kernicterus are
  1. salicylates
  2. sulfonamides
  3. furosemide
  4. gentamicin

Directions Summarized				
A	**B**	**C**	**D**	**E**
1,2,3	*1,3*	*2,4*	*4*	*All* are
only	only	only	only	correct

657. Variability of fetal heart rate may be influenced by
    1. gestational age
    2. vagal tone
    3. uterine blood flow
    4. magnesium sulfate

658. Iron deficiency anemia during pregnancy with a hemoglobin concentration of 9 to 11 g/dL is associated with
    1. no stainable iron in the bone marrow
    2. decreased serum iron
    3. elevated serum iron-binding capacity
    4. microcytosis

659. It is best to avoid as much as possible the administration of narcotics to parturient women, despite the availability of narcotic antagonists, because
    1. nalorphine may enhance respiratory depression not caused by narcotic drugs
    2. naloxone hydrochloride inhibits the analgesia produced by opioid narcotics
    3. levallorphan may enhance respiratory depression not caused by narcotic drugs
    4. naloxone hydrochloride does not reverse respiratory depression induced by opioid narcotics

660. Clinically, late deceleration of the fetal heart rate pattern may be associated with
    1. oxytocin infusion
    2. conduction anesthesia
    3. maternal hypotension
    4. head compression

661. The variable deceleration is a common fetal heart rate pattern in labor, felt to be caused by umbilical cord compression. The pattern is

   1. called variable because it does not look the same from one contraction to another
   2. not affected by maternal oxygen administration
   3. frequently present only intermittently
   4. associated with abrupt falls in heart rate to levels as low as 50 beats/min, but may be corrected by a change in maternal position

662. Pregnancy in women with sickle cell anemia is commonly associated with

   1. pulmonary disease
   2. perinatal death
   3. infection
   4. polycythemia

663. Multiple pregnancy is associated with an increased incidence of

   1. intrauterine growth retardation
   2. perinatal loss
   3. premature labor
   4. acute hydramnios

664. With external irradiation,

   1. as the energy of radiation increases, it becomes more penetrating
   2. as photons become more energetic, there is less lateral scatter of radiation
   3. the amount of radiation at any specific depth, compared to the surface dose, increases as the energy increases
   4. using cobalt teletherapy, good deep tissue doses are obtained because the surface tissues absorb minimal radiation

Directions Summarized

A	B	C	D	E
1,2,3	*1,3*	*2,4*	*4*	*All* are
only	only	only	only	correct

665. On colposcopy, the normal transformation zone is known
  1. to contain islands of columnar epithelium
  2. as the area between the original squamous epithelium and the columnar epithelium
  3. to contain metaplastic epithelium
  4. to contain atypical blood vessels

666. Adenomyosis is a benign disease of the uterus characterized by areas of endometrial glands and stroma within the myometrium. It is
  1. most commonly observed in women during the fifth and sixth decades of life
  2. found in approximately 20% of hysterectomy specimens
  3. not commonly found in association with endometriosis
  4. infrequently diagnosed correctly before surgery

667. Major problems in the management of ovarian cancer include
  1. major lymphatic drainage to the periaortic nodes
  2. metastases to the omentum, occurring early in the course of the disease
  3. the relative resistance of most tumors to radiation
  4. metastases to the omentum, often occurring before there is an obvious surface break

668. Laparoscopy is contraindicated in
  1. patients who are menstruating
  2. intestinal obstruction
  3. pelvic tuberculosis
  4. patients with hiatus hernia

669. Malignant lesions of the breast are the most common cancer in women. It is known that
1. 90% arise in the ductal system from long-standing hyperplasia of ductal cells
2. 1 of every 11 women is expected to develop breast cancer during her lifetime
3. stromal invasion ultimately occurs from in situ lesions
4. nodal metastasis is rare in 1 cm lesions

670. The preoperative evaluation of patients with cardiac disease is of the utmost importance, because it is well known that
1. patients with electrocardiographic evidence of ischemic heart disease are prone to have further EKG changes after surgery
2. patients with evidence of cardiac decompensation often may, in the postoperative period, demonstrate cardiac deterioration
3. the physical findings of a loud cardiac murmur or edema require thorough cardiac evaluation
4. the EKG may not be diagnostic of a prior myocardial infarction in a preoperative patient

671. Metabolic complications of intravenous hyperalimentation include
1. nonketotic acidosis
2. postinfusion hypoglycemia
3. dehydration
4. hyperkalemia

672. Changes in blood coagulation in the postoperative state include an increase in
1. circulating platelets
2. fibrinogen
3. platelet adhesiveness and aggregation
4. circulating fibrinolysin inhibitors

Directions Summarized				
A	**B**	**C**	**D**	**E**
1,2,3	*1,3*	*2,4*	*4*	*All* are
only	only	only	only	correct

673. Mucinous cystadenomas of the ovary
  1. are usually larger than serous cystadenomas
  2. secrete an acid mucopolysaccharide
  3. are bilateral in about 10% of cases
  4. that have papillary excrescences within the cyst should be considered malignant

674. Serous cystadenomas of the ovary are
  1. more frequently encountered in clinical practice than the mucinous variety
  2. present in one-third of all benign ovarian neoplasms
  3. commonly found to have papillary excrescences on the interior of the cyst wall
  4. never multilocular or parvilocular

675. Acute retention of urine may be caused by
  1. a pelvic hematocele
  2. an impacted myoma uteri
  3. a hematocolpos
  4. a retroverted gravid uterus

676. Urethral prolapse
  1. occurs in infants and children
  2. is often associated with continuous pain
  3. may cause acute urinary retention
  4. is often associated with prolonged catheterization after an anterior colporrhaphy

677. Imperforate hymen may be associated with
  1. hematocolpos
  2. dysuria
  3. hematometra
  4. urgency of micturition

678. Colposcopy can
1. differentiate between inflammatory atypia and neoplasia
2. not be relied on in many postmenopausal women
3. differentiate between invasive and noninvasive lesions
4. always furnish valuable information in all premenopausal women

679. Pulmonary embolism, although occasionally seen after surgical operations upon all parts of the body, is met with relatively greater frequency after abdominal and pelvic procedures than any others. It is recognized that
1. pulmonary embolism is asymptomatic in the vast majority of instances
2. the classic triad of chest pain, dyspnea, and hemoptysis is found in less than 5% of patients
3. the exhibition of symptoms depends on size, location, and number of emboli in the pulmonary arterial tree
4. when emboli obstruct more than 60% of the pulmonary arterial tree, there is an increase in right atrial and central venous pressure

680. In epidemiologic studies of cervical cancer, it has been shown that
1. celibacy protects the female from cervical carcinoma
2. divorced women have no higher incidence of the disease than matched controls
3. cervical cancer is rare when women experience coitus for the first time as late as age 27
4. prostitutes have the same incidence of the disease as matched controls

681. Endotoxic shock is a syndrome resulting from gram-negative bacteremia. In early septic shock,
1. the cardiac output is normal or elevated
2. the peripheral vascular resistance is reduced
3. systemic arterial hypoxemia is often found
4. the arterial PCO_2 tension is reduced

Directions Summarized				
A	**B**	**C**	**D**	**E**
1,2,3	*1,3*	*2,4*	*4*	*All* are
only	only	only	only	correct

682. In patients with anorexia nervosa, one notes
1. soft lanugo-type hair on the back
2. diabetes insipidus
3. bradycardia
4. hypercarotenemia

683. Prostacyclin is
1. a potent vasodilator
2. a prostaglandin
3. an inhibitor of platelet aggregation
4. a synthetic C prostaglandin

684. Sex hormone binding globulin (SHBG) is increased by
1. hyperthyroidism
2. growth hormone
3. pregnancy
4. testosterone

685. Von Willebrand disease is associated with
1. prolonged bleeding time
2. deficiency of Factor IX
3. prolonged partial thromboplastin time
4. Factor XI deficiency

686. Hematologic effects produced by steroidal antifertility agents include
1. increased sedimentation rate
2. decrease in prothrombin time
3. increased fibrinogen levels
4. decrease in clotting time

687. In the normal menstrual cycle,
   1. the LH surge precedes ovulation by 34 to 36 hours
   2. postovulatory progesterone is secreted primarily from the theca cells
   3. progesterone secretion begins in the late follicular phase
   4. FSH increases due to a positive feedback with estradiol

688. Successful treatment of breast cancer depends on early detection. Initial findings include
   1. palpable mass
   2. nipple discharge
   3. nipple retraction
   4. "positive" mammography

689. Adenomyosis and endometriosis have similar clinical presentations. They have in common
   1. pelvic pain and worsening dysmenorrhea
   2. large, tender uterus
   3. abnormal uterine bleeding
   4. cyclic hematochezia

690. Pelvic inflammatory disease is used to describe varying stages of salpingitis to pelvic peritonitis. The clinical picture includes
   1. abdominal pain starting in the pelvis
   2. leucocytosis with left shift
   3. high fever and pelvic tenderness on examination
   4. elevated amylase

691. The diagnosis of ectopic pregnancy
   1. is suspected in patients who present with pain, bleeding, and an adnexal mass
   2. rupture can be diagnosed via culdocentesis
   3. may be confirmed by quantitative beta HCG and ultrasound
   4. can be ruled out by identifying an intrauterine pregnancy with ultrasound

Directions Summarized

A	B	C	D	E
1,2,3	*1,3*	*2,4*	*4*	*All* are
only	only	only	only	correct

692. Infertility
   1. is rare in the United States
   2. may be associated with Asherman's syndrome
   3. is usually caused by failure of capacitation
   4. can usually be corrected

693. Amenorrhea is a fairly common complaint estimated to occur in 2% to 5% of women. Assuming pregnancy has been ruled out,
   1. a young, thin distance runner with secondary amenorrhea can simply be reassured
   2. a young amenorrheic woman who also complains of headache and breast discharge should be further evaluated
   3. an obese, hypertensive diabetic 45-year-old should have an endometrial biopsy
   4. in an estrogenic, ovulatory woman who had a curettage for spontaneous incomplete abortion 10 months ago, an HSG may be diagnostic

694. When a patient is seen with continuous bleeding for the past three weeks, it is important to determine if
   1. she or any family members are "free bleeders"
   2. she is or could be pregnant
   3. she wishes to get pregnant soon
   4. she has any adverse reactions to estrogens and/or progestins

695. The initial laboratory evaluation of hirsutism should include
   1. serum testosterone
   2. serum prolactin
   3. serum DHAS (dehydroepiandrosterone sulfate)
   4. serum 17 OHP (17 hydroxyprogesterone)

696. Which of the following statements concerning late onset adrenal hyperplasia is(are) true?
   1. It is diagnosed by 17 hydroxyprogesterone assay
   2. Pregnant couples with this condition should receive genetic counseling
   3. These patients may be subjected to cortisol deficiency during severe stress
   4. It is the most common autosomal recessive disorder

697. Which of the following statements regarding androgen-producing tumors is(are) true?
   1. Causative factor in 25% of cases of hirsutism
   2. Associated with testosterone levels of 200 mg/dL or greater
   3. Functioning tumors are usually not palpable
   4. Associated with rapidly progressing masculinization

698. Which of the following statements concerning the treatment of hirsutism is(are) true?
   1. Androgen production in the hirsute woman is usually an LH-dependent process
   2. Progestational agents and oral contraceptives are useful because of their potent negative feedback actions on LH
   3. The estrogen components in oral contraceptives are useful because they increase SHBG, and therefore decrease free testosterone
   4. Combined treatment with electrolysis is not recommended until after six months of hormonal suppression

699. Which of the following statements concerning spironolactone is(are) true?
   1. It inhibits ovarian biosynthesis of androgens
   2. It inhibits adrenal biosynthesis of androgens
   3. It competes for the androgen receptor in the hair follicle
   4. It directly inhibits 5-α-reductase activity

Directions Summarized				
A	B	C	D	E
1,2,3	*1,3*	*2,4*	*4*	*All* are
only	only	only	only	correct

700. Which of the following regarding Müllerian agenesis is(are) true?
1. Karyotype 46XX
2. Normal sexual hair
3. Other abnormalities are frequent
4. X-linked recessive disorder

Explanatory Answers

585. C. A neonatal death refers to death of a live-born infant within 28 days of birth. The birth rate is the total number of births per 1000 population. The neonatal death rate is the number of deaths within the first 28 days of life per 1000 live births. This may be divided into the early neonatal deaths, in the first 7 days of life, and late neonatal deaths, after day 7 but before day 29. In international statistics, the neonatal death rate is quoted. The stillbirth rate is the number of stillborn infants per 1000 total births, live and stillborn. (Ref. 1, p. 2)

586. D. Strands of K-cells migrate into the corpus luteum from the theca. They become more prominent and contain lipid as well as high concentrations of alkaline phosphatase. (Ref. 1, pp. 916–918)

587. E. The syncitium is derived from the cytotrophoblast. This process, which results in an increasing cell mass capable of significant metabolic function as pregnancy progresses, is continuous. No mitotic figures are seen suggesting that it is a mitotic end stage. There is an abundant endoplasmic reticulum with well developed Golgi complexes. The syncytiotrophoblast is a source of placental enzymes. The cytotrophoblast is relatively simple. (Ref. 1, pp. 48–53)

588. D. HPL has several ill-defined multiple metabolic effects, including lipolysis and glucose homeostasis. It has been detected in sera from persons with malignancies originating in the trophoblast, gonad, lung, liver, and hemopoietic system. (Ref. 1, pp. 70–71)

589. A. The glycerophospholipids (surfactant) are important in the maturation of fetal lungs, allowing aeration with the first breath and the prevention of respiratory distress. Approximately 80% of the phospholipids are lecithins, the largest fraction being dipalmitoylphosphatidylcholine. On the other hand, phosphatidylglycerol, although only comprising about 9% of the total, appears to be a factor that stabilizes the surfactant and is probably important in preventing subsequent respiratory distress. (Ref. 1, pp. 107–109)

590. C. In normal pregnancy, there is an increase in the blood volume, maternal body weight, and basal metabolic rate. Vascular resistance and blood pressure fall. To compensate, the stroke volume increases, resulting in an increase in cardiac volume by 10% and the resting pulse rate increases by 10 to 15 beats/min. As the intra-abdominal mass increases, the diaphragm is splinted upwards and the rib cage expanded, resulting in displacement of the heart upwards and to the left with some axis deviation, which shows in the electrocardiogram. (Ref. 1, p. 144)

591. C. Engagement is defined as the passage of the largest diameter of the presenting part through the pelvic inlet. When a breech is presenting, the widest diameter is the bitrochanteric and with a cephalic presentation it is the biparietal. Fixation of the presenting part in the inlet can occur even when true disproportion is present and is not an index of engagement. In primagravidae, the presenting part is normally engaged at or about 36 weeks gestation; engagement is not usual in multiparae until the onset of labor. (Ref. 1, pp. 169–170)

592. C. Closed neural tube defects, eg, hydrocephaly, are not associated with raised α-fetoprotein. Included in the conditions in which raised maternal and/or fetal α-fetoprotein is found are open neural tube defects, congenital nephrosis, bladder neck obstruction, esophageal and duodenal atresia, exomphalos, sacrococcygeal tumors, pilonidal sinus, Turner's syndrome (45XO), fetal death, fetal blood in amniotic fluid, and fetomaternal hemorrhage. (Ref. 1, pp. 584–586)

593. D. The latent phase of labor precedes the active phase. (Ref. 1, p. 220)

594. C. In a hydatidiform mole, the chorionic villi are hydropic and contain no fetal vessels. The embryo dies at an early age. Hyperplasia of the cytotrophoblast, which produces HCG, is present. The syncytium merely stores HCG. In the syncytium, numerous large nuclear masses are noted and scant cytoplasm. (Ref. 1, p. 541)

595. B. Polyhydramnios, or hydramnios, as opposed to oligohydramnios, may be chronic or acute. The former is more common than the latter. Acute polyhydramnios is associated with monozygotic twins; chronic polyhydramnios is associated with maternal diabetes, rhesus isoimmunization, and fetal anomalies, often with open central nervous system defects. However, it should be remembered that the majority of pregnancies complicated by too much amniotic fluid are associated with apparently normal neonates. The increase in the perinatal mortality is due to complications of delivery, abnormal presentations, and prolapse of the cord, as well as the associated maternal and fetal diseases. (Ref. 1, pp. 554–558)

596. C. Preeclampsia is defined as the development of hypertension and albuminuria or edema or both during pregnancy. When this condition is complicated by the occurrence of convulsions, it is described as eclampsia. (Ref. 1, pp. 654–655)

597. E. Acute pyelonephritis complicating pregnancy occurs more commonly as pregnancy advances and during the puerperium, affecting about 2% of patients. Symptoms usually have a sudden onset and include bladder irritability, hematuria, fever, chills, anorexia, nausea and vomiting, and an aching pain in one or both lumbar regions. Causative microorganisms come from a wide spectrum; the most common isolate is *E. coli*. There is no suggestion of a deficiency in the autoimmune system being associated with pyelonephritis. (Ref. 1, pp. 809–810)

598. A. Studies have suggested that iron 30 mg daily as a simple salt such as ferrous gluconate, sulfate, or fumerate, taken regularly once each day throughout the latter half of pregnancy, should provide sufficient iron to meet the requirements of pregnancy and to protect any preexisting iron stores. The pregnant woman may benefit from 60 to 100 mg of iron per day if she is large, has twin fetuses, is late in pregnancy or takes iron irregularly or her hemoglobin level is somewhat depressed. The woman who is overtly anemic from iron deficiency responds well to 200 mg of iron per day in divided doses. (Ref. 1, pp. 264, 780–782)

599. B. During pregnancy, maternal viremia with rubella will result in a high incidence of abortion with congenital deformities. The neonate may have lesions of the eyes, central nervous system, cardiovascular system, respiratory system, and hematologic system. Infants born with congenital rubella may be infective. The rubella virus may induce chromosomal anomalies and be diabetogenic. (Ref. 1, p. 616)

600. B. Breast engorgement generally precedes infection, which is diffuse but later localizes. Symptoms are usually localized pain, swelling, and redness, often associated with chills or rigors. *S. aureus* commonly gains entry through cracked skin or nipples. Most important is prevention by handwashing by the mother and attendants. Breast care includes keeping skin clean and supple. Samples of breast milk should be obtained before antibiotic therapy is started. Suppression of lactation is normally indicated. (Ref. 1, p. 485)

601. E. Colonization of the female genital tract with group B, β-hemolytic streptococci occurs with increasing frequency as pregnancy progresses, being found in as many as 40% of pregnant women in the third trimester. Iatrogenic introduction does not appear to be a common program. Neonatal colonization is relatively common; but neonatal disease is infrequent. Early onset disease, characterized by septicemia, has a high mortality rate. Late onset disease, often with meningitis, is less than that for early onset disease. Treatment at birth with penicillin in those cases where the infection can be expected, after prolonged spontaneous rupture of membranes, for example, appears to be effective in lessening the incidence of neonatal disease. (Refs. 1, pp. 614–615)

602. E. The incidence of Down syndrome increases with maternal age. The most frequently found abnormality is trisomy 21, although trisomy 18 appears to be a more lethal chromosomal rearrangement. (Ref. 1, pp. 570–571)

603. E. Disturbances of body image are sometimes manifested as anorexia nervosa. This is a disease that is most common between the ages of 11 and 21, and overwhelmingly affects females. There is a pathological refusal, or inability, to recognize hunger. Most physical and laboratory tests are normal. (Ref. 3, p. 197)

604. B. The usual karyotype of hydatidiform mole is 46XX, with the chromosomes completely of paternal origin—the latter phenomenon is referred to as androgenesis. The ovum is fertilized in such instances by a haploid sperm which duplicates its own chromosomes after meiosis. The original ovum chromosomes are either inactivated or absent. (Ref. 1, p. 541)

605. D. The most frequently encountered ovarian tumor in the young reproductive years is the dermoid cyst. This consists of all three germ cell layers and, as such, may contain any variety of tissue elements. They are particularly liable to undergo torsion because they are mobile and usually on a long pedicle. Malignant change is uncommon. (Ref. 2, p. 832)

606. D. Müllerian duct regression factor is produced by the Sertoli cells of the seminiferous tubules. The seminiferous tubules appear in the fetal gonad before the Leydig cells, which are the cellular site of origin of testosterone. (Ref. 1, p. 120)

607. A. In testicular feminization, there is little or no response to androgen. These patients have no uterus, fallopian tubes, or Wolffian duct structures, but have a short, blind-ending vagina and breast development. (Ref. 1, p. 122)

608. C. Renal agenesis and the associated anomalies are commonly referred to as Potter syndrome. The skin is loose and the hands often seem large. Other features include prominent epicanthal folds, large low-set ears, flattened nose, and, in about one-third, cardiac anomaly. (Ref. 1, p. 577)

609. E. There are several drugs, such as diazepam, morphine, meperidine, atropine, scopolamine, and phenothiazines, which, when administered to the mother, are associated with ablation or a marked reduction in beat-to-beat variation in fetal heart rate. (Ref. 1, p. 299)

610. C. The most sensitive marker for increased androgen production is hirsutism. This is followed, in order, by acne and increased oiliness of the skin, menstrual irregularity, increased libido, clitoromegaly, and, finally, masculinization. Masculinization and virilization are terms reserved for extreme androgen production, usually, but not always, associated with a tumor. (Ref. 3, p. 239)

611. C. Laparoscopy and ovarian biopsy are not indicated procedures in the evaluation of hirsutism. (Ref. 3, p. 258)

612. A. A sinusoidal fetal heart rate pattern is a regular, smooth sine wavelike baseline with absence of beat-to-beat variability. The pattern was originally described in severely affected Rh-isoimmunized fetuses. It has since been observed in association with severe anemia of any etiology. It is also noted in about 50% of fetuses following administration of alphaprodine (Nisentil) to the mother for relief of discomfort during labor. A sinusoidal pattern has also been associated with maternal administration of meperidine, with reversal of this pattern when naloxone was administered. Butophanol (Stadol) has also been reported to induce sinusoidal patterns in the fetus. (Ref. 1, p. 299)

613. A. The human ovary has the enzymatic capability of producing dehydroisoandrosterone, androstenedione, and testosterone. Androstenedione levels increase sharply just prior to ovulation, fall in the early luteal phase, and rise again in the late luteal phase. The adrenals are the other source of androstenedione. Dehydroisoandrosterone is secreted principally by the adrenal; testosterone is derived mainly from peripheral conversion. Androstenedione represents an important prehormone for the peripheral conversion to androgens and estrogens. (Ref. 1, pp. 911–912)

614. E. The development of the male fetus is similar to that of the female except the germ cells locate in the medulla, rather than the cortex, as in the female. The meiotic divisions result in spermatocytes, each with an equal share of the genetic material; whereas in the female, genetic material is discarded as polar bodies. (Ref. 1, p. 888)

615. E. Placental transfer is a complex physiological process resulting in the supply of oxygen and nutrients and the removal of carbon dioxide and other waste products. There are many factors that can influence the exchange between mother and fetus. These include all of the above, the degree of protein binding, the physical properties of the barrier, and the capacity of the biochemical mechanism for active transfer. (Ref. 1, p. 95)

616. E. In utero, the high pulmonary vascular resistance, resistance to flow through the ductus arteriosus associated with the low resistance umbilicoplacental circulation, allows preferential flow of oxygenated blood through the ductus venosus directly into the inferior vena cava, through the foramen ovale, into the left atrium. Thus, in an environment where the PO_2 of fetal blood is at best poor, the vital organs receive an adequately oxygenated blood flow. With expansion and oxygenation of the lungs after birth, a fall in pulmonary vascular resistance, there is a reorganization towards an adult circulation. Variations in oxygen tension, together with changes in prostaglandin metabolism, appear to be the main factors associated with the vascular changes. (Ref. 1, pp. 99–101)

617. E. Development of the male depends upon the presence of a Y chromosome. Sertoli cells of the testis produce Müllerian duct inhibiting factor and Müllerian duct regression. Leydig cells produce testosterone, which acts upon the Wolffian ducts to produce the epididymis, vas deferens, and seminal vesicles, and dihydro-testosterone, which influences the development of the penis and scrotum. All intrauterine ambiguous sexual development results from abnormal androgen representation. (Ref. 1, pp. 119–120)

618. E. Resuscitative measures, beyond oropharyngeal suction and superficial stimulation, may be unsuccessful for a number of reasons. Equipment failures, damaged resuscitation bag, or faulty laryngo-scope, an unsuitable environment, poor intubation technique or congenital abnormalities, inefficient operation of equipment, and diagnostic failures all may account for poor results from the resuscitative effects. (Ref. 1, p. 241)

619. B. Consumptive coagulopathy, or intravascular coagulation defects, and renal failure may be directly associated with abruptio placentae. These complications must be searched out in every case where placental separation occurs. (Ref. 1, p. 706)

620. C. Carcinoma of cervix is overwhelmingly (90%) of the squamous variety; the majority of lesions arising at the squamoco-lumnar junction at or near the external os. Adenosquamous lesions carry the same prognosis as either squamous carcinoma or adenocarcinoma. (Ref. 2, p. 682)

621. A. Acute pelvic inflammatory disease represents a diagnostic problem. The differential diagnosis for acute pelvic when the disease is limited to the pelvis must include consideration of infections

270 / Clinical Sciences

affecting all the pelvic organs. When peritonitis becomes a complication, then adjoining organ systems must be considered, ie, lower lobe pneumonia. (Ref. 2, p. 507)

622. E. Intrauterine growth retardation (IUGR), small for gestational age, is multifactorial, such factors appearing singly or in combination in an individual pregnancy. These factors include maternal age, chronic hypoxia, genetic factors and chromosomal abnormalities, intrauterine infections, and placental anomalies. Recognition of intrauterine growth retardation is a most important aspect of prenatal care. In general, asymmetric IUGR does not carry such a high perinatal morbidity and mortality as does symmetric IUGR. (Ref. 1, pp. 765–766)

623. C. Asymptomatic bacteriuria occurs in 5% to 7% of all pregnant women and is caused predominantly by *E. coli*. (Ref. 1, pp. 808–809)

624. B. The late follicular phase rise in estradiol is probably responsible for setting off the LH surge at midcycle. Abnormal levels of androgens probably suppress the cyclic center. However, exogenous estrogens probably influence gonadotropin release. FSH induces an increase in LH receptor sites, and it is probable that a midcycle surge is necessary for the LH surge to produce steroidogenesis and a normal corpus luteum. (Ref. 3, p. 97)

625. B. The characteristics of Turner's syndrome (karyotype XO) are short stature, webbed neck, shield chest, and hypergonadotropic hypoestrogenic amenorrhea. A female child with bilateral inguinal hernia should be suspected of having the testicular feminizing syndrome, testes, and an XY karyotype. The resistant ovary syndrome is characterized by elevated gonadotropins, ovarian follicles, and amenorrhea. (Ref. 3, p. 401)

626. E. Pruritus vulvae is the most common gynecologic symptom bringing a woman to the doctor's office. The gynecologist must be aware that the symptom of pruritus vulvae is a representation of numerous disease processes. (Ref. 4, p. 1015)

627. C. Before a sperm can fertilize an oocyte, it must undergo capacitation, which is the removal of a protective coating from the head of the sperm. Following capacitation, the sperm undergoes an acrosome reaction, during which time small perforations are produced in the acrosome wall. Such small perforations allow the escape of enzymes that digest a path for the sperm through the corona radiata and zona pellucida. The primary spermatocyte undergoes the first meiotic division; the secondary spermatocytes undergo a second meiotic division. One primary spermatocyte results in the formation of four sperms, whereas only one mature oocyte results from maturation of a primary oocyte. (Ref. 3, p. 500)

628. C. The umbilical vessels are derived from the blood vessels of the allantois. Likewise, after birth the allantois forms the urachus, which is the median umbilical ligament. Blood development occurs in the walls of the yolk sac from the third week and continues until hemopoietic activity commences in the embryonic liver during the fifth week. During the fourth week, the dorsal part of the yolk sac forms the embryonic endodermal tube, the primitive gut, which gives rise to the epithelium of the digestive and respiratory tracts. By 12 weeks the small yolk sac lies in between the chorionic sac and the amnion. The yolk sac usually detaches from the gut by the end of the fifth week, but in about 2% of adults the proximal intra-abdominal part of the yolk sac persists as a diverticulum of the ileum known as Meckel's diverticulum. (Ref. 1, pp. 46, 64)

629. C. The round ligaments arise somewhat below and anterior to the fallopian tube. Each passes outward and downward through the inguinal canal, to be inserted into the upper portion of the labium majus. It consists of smooth muscle cells that are continuous with those of the uterine wall. The round ligament together with the ovarian ligament is the homologue of the gubernaculum testis in the male. (Ref. 1, p. 882)

630. B. The medial femoral circumflex artery anastomoses with the obturator and inferior gluteal arteries, which arise from the hypogastric artery. The lateral femoral circumflex artery anastomoses with the superior gluteal and iliolumbar arteries, which arise from the hypogastric artery. (Ref. 5, p. 53)

631. C. Anteriorly and medially, the hypogastric artery is covered by the peritoneum with varying amounts of fat and areolar tissue. Under cover of the peritoneum, the ureter frequently descends along the anterior border of the artery. Immediately posterior to the artery are the hypogastric vein and the origin of the common iliac vein. The external iliac vein lies adjacent to the artery. (Ref. 5, p. 49)

632. E. The cardinal ligaments anchor the cervix and vaginal vault, and counteract the weakening of the pelvic diaphragm caused by its being pierced by the vagina. The cardinal ligaments and the uterosacral ligaments are condensations of the endopelvic fascia. (Ref. 4, p. 39)

633. A. The gonads are derived from three sources: the coelomic epithelium, the underlying mesenchyme, and the primordial germ cells. Gonadal development is first indicated during the fifth week, when a thickened area of coelomic epithelium develops on the medial aspect of the urogenital ridge. The primordial germ cells are visible in the fourth week in the wall of the yolk sac, near the origin of the allantois. The primordial germ cells migrate during the sixth week to be incorporated in the primary sex cords. (Ref. 1, p. 888)

634. E. Delayed puberty is a rare condition in females. Only 1% of all girls will not have had menarche by age of 18 years. Causes include constitution, malnutrition, juvenile diabetes mellitus, gonadal dysgenesis, gonadotropin deficiency, and anosomia, as in Kallmann's syndrome, as well as anorexia nervosa. About 20% of patients with sickle cell disease have delayed puberty. (Ref. 3, p. 428)

635. B. The catecholamines are formed in the brain from tyrosine. Serotonin is an indolamine formed from tryptophan, and at present is the only indolamine associated with the release of prolactin. (Ref. 3, p. 51)

636. C. Prolactin secretion is inhibited by serotonin antagonists such as methysergide, L-dopa, and certain ergot alkaloids. Thyrotropin-releasing hormone has a direct stimulatory effect on the pituitary, causing it to release prolactin. (Ref. 3, p. 59)

637. E. It is recommended that estrogens be given to all menopausal women indefinitely, except when specifically contraindicated. It alters the lipoprotein profile, producing a cardioprotective effect, relieves vasomotor symptoms, helps prevent the progression of bone loss leading to osteoporotic fractures, and prevents or reverses genitourinary atrophy. (Ref. 3, p. 134)

638. E. Other contraindications are known or suspected estrogen-dependent neoplasia (eg, breast and endometrial cancer), pregnancy, and smoking over age 35. Relative contraindications include migraine headache, leiomyomas, epilepsy, uncontrolled hypertension, gall bladder disease, diabetes and SS and SC hemoglobinopathies. (Ref. 3, p. 481)

639. B. The normal menstrual cycle begins with low levels of estradiol and progesterone. FSH increases estradiol secretion, which in turn suppresses FSH secretion. The sharp rise in estrogen in the presence of FSH increases LH and LH receptors in the granulosa cells, which begin to secrete progesterone. This is associated with a late follicular phase increase in FSH, which then falls to low levels in the luteal phase. After the surge, LH falls to low levels, but continues to stimulate the corpus luteum production of estradiol and progesterone. Androstenedione and testosterone are at higher levels at midcycle than at any other phase of the cycle. (Ref. 3, p. 97)

640. B. Testosterone, androstenedione, dehydroepiandrosterone, and its sulfate conjugate are quantitatively the principal androgens involved in hyperandrogenic states in women. 19-Nortestosterone derivatives are gestagens used in oral contraceptive formulations. Androsterone and etiocholanolone are 17-ketosteroids. Testosterone is about tenfold as potent as androstenedione and about twenty times as potent as dehydroepiandrosterone. (Ref. 3, p. 220)

641. A. Patients with sex chromatin-negative 45XO gonadal dysgenesis often have micrognathia, high arched palate, dental abnormalities, broad shieldlike chest, hypoplastic areolae, short neck with low hairline, recurrent otitis media, short fourth metacarpals, and cubitus valgus. Cardiovascular and renal anomalies are also frequently noted. The biphasic pattern of gonadotropin secretion in normal childhood is exaggerated in patients with gonadal dysgenesis because of the absence of gonadal steroids to exert negative feedback. (Ref. 3, p. 401)

642. B. The fetal zone of the fetal adrenal cortex of normal fetuses secretes large amounts of dehydroepiandrosterone sulfate, estimated at 100 to 200 mg/d. The fetal adrenal also produces the 16-hydroxylated derivative of this C19 steroid, although most of the 16-hydroxylation occurs in the fetal liver. The resultant 16-hydroxylated substrate is then used by the placental trophoblast for estriol biosynthesis. An identical process occurs on the maternal side, where adrenal dehydroepiandrosterone sulfate undergoes 16-hydroxylation in the maternal liver, and is then available to the trophoblast of the placenta for the formation of estriol. During pregnancy, however, the maternal adrenal only secretes 8 to 10 mg per day of dehydroepiandrosterone sulfate, and consequently the fetal zone of the fetal adrenal gland is the principal site for the supply of the C19 precursor for estriol synthesis by the placenta. This latter fact has established the evaluation of maternal urinary estriol concentrations as a significant parameter in the evaluation of the integrity of the fetoplacental unit and fetal well-being. (Ref. 1, p. 72)

643. A. Accelerated biotransformation of estrogens to less active metabolites may be responsible for breakthrough bleeding, breakthrough ovulation, and pregnancy in women on antifertility steroid preparations. Rifampicin, which is used in tuberculosis therapy, has this property. Other drugs that have been implicated in affecting the metabolism of oral contraceptives in this way are anticonvulsants (eg, phenytoin, barbiturates), phenylbutazone, and benzodiazepines. Other drugs that interact with OCs are phenothiazides, anticoagulants and certain antihypertensives (eg, guanethidine). From the practical standpoint, a physician should advise alternative contraception to patients on the above drugs. (Ref. 3, p. 482)

644. A. The most common form of sexual precocity in females is idiopathic or constitutional. Albright's syndrome, known as polyostotic fibrous dysplasia, consists of multiple, disseminated cystic bone lesions, café au lait skin areas, and sexual precocity. Encephalitis, Von Recklinghausen's disease, intracranial tumors, and hypothyroidism may be associated with precocious puberty. (Ref. 3, p. 418)

645. E. The most frequent side effect of clomiphene citrate therapy, which occurs in up to 10% of patients, is the occurrence of vasomotor disturbances. Abdominal distention and discomfort, mammary discomfort, headache, visual disturbances (particularly blurred vision), nausea and vomiting, and dryness or loss of hair are also encountered as transient side effects. Significant ovarian enlargement is encountered when larger dosage schedules are used, but is an infrequent side effect when a five-day course of clomiphene in a daily dosage of 100 mg is prescribed. (Ref. 3, p. 591)

646. E. In the differential diagnosis of hyperprolactinemia, one must include prolactin-secreting pituitary tumor, hypothyroidism, renal failure, stress, chest wall diseases, spinal cord lesions, and drugs such as phenothiazines, reserpine, tricylcic antidepressants and α-methyldopa. (Ref. 3, p. 291)

647. C. Congenital adrenal hyperplasia associated with 11 β-hydroxylase deficiency results in excess deoxycorticosterone production, salt retention, virilization, hypertension, and hypokalemia. Such individuals are unable to produce aldosterone, corticosterone, and hydrocortisone. Congenital adrenal hyperplasia very frequently occurs in siblings; affected siblings also demonstrate the same enzyme defect, but not always to the same extent. (Ref. 4, p. 139)

648. C. Class A is a chemical diabetes controlled with diet. All other classes require insulin. Class B has its onset at greater than age 20 and duration less than 10 years. Class C onset occurs between 10 and 19, or duration is between 10 and 19 years, with no evidence of vascular disease. Class D has an onset before age 10 or duration more than 20 years. Other classes refer to end-organ derangement: F, nephropathy; R, proliferative retinopathy; H, heart disease. (Ref. 1, p. 816)

649. E. Epidural analgesia is produced by injecting a local anesthetic agent into the epidural space. Uterine contraction sensation is blocked by anesthetizing the spinal segments T10-L2. Delivery comfort is produced by blocking S2-S4. Patients should be prehydrated with 500 to 1000 mL of isotonic solution prior to the block to prevent hypotension from the sympathetic blockade. Epidurals should not be used in the presence of maternal hemorrhage, infection, anticoagulant therapy, or bleeding disorders. Direct intravascular injection of local anesthetic may result in grand mal seizures, cardiopulmonary arrest, and fetal bradycardia. (Ref. 4, p. 671)

650. E. Fetal IUGR is associated with thin or immature maternal body habitus and low weight gain, cyanotic heart disease, malabsorption syndromes, smoking, and hypertension. It is also associated with various fetal trisomies and other chromosomal anomalies, organ system anomalies, and infection with any of the TORCH agents. (Ref. 4, p. 467)

651. E. Excessive doses of vitamin K analogs may be associated with hyperbilirubinemia. Vitamin K_1 seems to be much safer than some of its synthetic analogs such as Mephyton and Synkayvite. Sepsis contributes to kernicterus by an unclear mechanism. Both hypothermia and hypoglycemia predispose the infant to kernicterus by elevating nonsterified fatty acids, which compete with bilirubin for albumin-binding sites. (Ref. 1, p. 608)

652. A. Hurler's syndrome, Tay-Sachs disease, and Pompe's disease are inherited as autosomal-recessive disorders in which the possibility of an abnormal child from normal carrier parents is 25%; consequently, amniocentesis is indicated. Lesch-Nyhan syndrome and Duchenne-type muscular dystrophy are X-linked conditions in which 50% of male infants of carrier mothers are affected. In such instances, amniocentesis allows determination of fetal sex. (Ref. 4, p. 39)

653. C. The oral glucose tolerance test in pregnancy is altered by a small elevation in blood glucose in the postabsorptive phase and a delay in returning to normal. Duration of the gestation influences the glucose curve; the results of repeated tests in the same individual may vary. (Ref. 4, p. 546)

654. E. Glucose-6-phosphate dehydrogenase deficiency is a relatively common hereditary disorder present in about 10% of the black population in the United States. In populations with a high incidence of this disorder, agents capable of inducing hemolysis, such as sulfonamides, nitrofurantoins, salicylates, and water-soluble analogs of vitamin K, are to be avoided. Perinatal risks include fetal hydrops, neonatal jaundice, and bilirubin encephalopathy. The greatest maternal risk is the high incidence of urinary tract infections. Anemia is also frequent in such women during pregnancy because of hemolysis. The presence of a relatively normal serum iron during pregnancy, together with an anemic state, is an indication for a screening test of glucose-6-phosphate dehydrogenase deficiency. (Ref. 4, p. 536)

655. A. Ritodrine produces less cardiovascular effects than its analog isoxsuprine. When administered in an intravenous dosage in excess of 400 μg/min, maternal tachycardia frequently occurs. The maximum dose of this drug is consequently limited by the elevated maternal pulse rate. (Ref. 4, p. 686)

656. E. Bilirubin is readily bound to serum albumin, which makes it less liable to affect the cells within the central nervous system of the neonate. However, in prematurity, the level of albumin is low. Such agents as sulfonamides, salicylates, furosemide, and gentamicin compete with bilirubin for attachment to albumin and, as a result, allow toxic amounts of unconjugated and non-protein-bound bilirubin to reach the central nervous system. Some such mechanism accounts for the fact that kernicterus may occur at different levels of serum bilirubin. It should be realized that the more premature the infant, the lower the bilirubin level necessary to produce kernicterus; and the more damaged the brain cells by hypoxia, the more readily they are predisposed to this type of damage. (Ref. 1, p. 608)

657. E. Variability has been shown to be an important commentary of fetal heart rate pattern and has been adequately shown to correlate with fetal Apgar scores and pH values. Decreased variability on its own is almost never indicative of asphyxia, unless it is accompanied by decelerations. Variability of fetal heart rate may be influenced by gestational age, vagal tone, uterine blood flow, and drugs. Compounds such as meperidine, magnesium sulfate, and diazepam may ablate beat-to-beat variation. The normal fetal heart rate pattern has the following features: (1) a rate within the range of 120 to 160 beats/min; (2) no change in rate during a uterine contraction; and (3) variability (beat-to-beat variation) of five or more beats/min. A normal pattern is reliable evidence that the fetus is in good condition, and it is unnecessary to investigate the fetus further by pH management so long as this pattern persists. (Ref. 1, p. 296)

658. A. A hemoglobin concentration of 9 to 11 g/dL is usually not accompanied by obvious morphologic changes in the circulating erythrocytes. Microcytosis and erythrocytic hypochromia are not observed. The serum iron, however, is below normal and no stainable iron is present in the bone marrow. The serum iron-binding capacity is increased, but is normally elevated during pregnancy in the absence of iron deficiency, and is of little diagnostic value. The initial evaluation of a gravid patient with moderate anemia should include such investigations as hemoglobin, hematocrit, peripheral blood smear, serum iron concentration, serum ferritin, and, if the patient is black, a sickle cell preparation. A bone marrow preparation is not necessary. (Ref. 1, p. 781)

659. A. Naloxone hydrochlorine (Narcan) is a narcotic antagonist capable of reversing to some degree the respiratory depression induced by opioid narcotics, but it also inhibits the analgesia and euphoria produced by the narcotic. Levallorphan and nalorphine enhance respiratory depression not caused by narcotic drugs. Although narcotic antagonists may help relieve respiratory depression from opioid drugs in the neonate, it is best to avoid as much as possible the administration of narcotics to the parturient when they might cause respiratory depression in the neonate. (Ref. 1, p. 329)

660. A. Clinically, late deceleration is associated with conditions that cause a reduction in the intervillous space blood flow. Consequently, maternal hypotension, uterine hypertonus, and the uteroplacental pathology encountered in hypertension, eclampsia, diabetes mellitus, and postdate gestations can all result in decreased intervillous space blood flow. The supine hypotensive syndrome, conduction anesthesia, and oxytocin infusion are the most common causes. In fact, the combination of oxytocin infusion and conduction anesthesia may produce late deceleration in about 40% of patients. The most ominous form of late deceleration is the one that is shallow, associated with baseline tachycardia and loss of beat-to-beat variability. (Ref. 4, p. 785)

661. E. The variable deceleration pattern shows a nonuniform or variable shape and temporal relationship to the associated uterine contraction. Characteristically, they can be made less severe with changes in maternal posture, in accord with the etiology of umbilical cord compression. For the most part, this is an innocuous pattern similar to the vagal, type I, deceleration and is not usually associated with hypoxia. However, prolonged, deep, and frequently recurrent variables can be equated with late decelerations, and may result in fetal hypoxia and acidoses. (Ref. 4, p. 783)

662. A. About half of the gravid patients with sickle cell anemia will have either an abortion, a stillbirth, or a neonatal death. They tend to have more frequent episodes of pain. Pulmonary disease and infections are more common in these women. Although the platelet count may be increased, polycythemia is never present. (Ref. 1, p. 785)

663. E. Acute hydramnios is associated with monozygotic twins. Both intrauterine growth retardation and premature delivery are associated with multiple pregnancy, and the larger the number of fetuses, the greater is the liability for these conditions to occur. If multiple pregnancy is not suspected, there may be a delay in delivery of the second twin, with an increased possibility of traumatic delivery, hypoxia, and hemorrhage. (Ref. 1, p. 629)

664. A. The energy and penetrating power of ionizing radiation increase as the photon wavelength diminishes. As photons become more energetic, they travel a greater distance into absorbing tissues. With supervoltage radiation, the maximum ionization occurs below the level of the epidermis, and there is less lateral scatter of radiation. With cobalt teletherapy, maximum ionization occurs about 5 mm below the surface, while the surface dose is about 40% of the maximum. Clinical observations have shown that as radiation energy increases, similar tumor effects can be produced with less damage to adjacent normal tissues. (Ref. 4, p. 1201)

665. A. Original squamous epithelium refers to the smooth pink epithelium originally established on the cervix and vagina. The transformation zone is the area between the original squamous epithelium and the columnar epithelium, in which metaplastic epithelium in varying degrees of maturity is observed. The transformation zone contains columnar epithelium, gland openings, and nabothian cysts. Atypical blood vessels are not present, as these are a feature of an atypical transformation zone, as is a mosaic pattern, and both are often seen in cervical dysplasia and carcinoma in situ. (Ref. 4, p. 1059)

666. E. The classic manifestations of adenomyosis are progressively heavy menstrual bleeding, dysmenorrhea, and a tender, gradually enlarging uterus. All or part of the clinical picture of adenomyosis occurs with other more readily diagnosed conditions. Myomas are a much more common cause of uterine enlargement, and endometrial hyperplasia is a more common cause of menorrhagia. Marked endometrial hyperplasia can cause uterine enlargement. Secondary dysmenorrhea, although a common complaint of women with adenomyosis, is more common in patients without pelvic disease and in females with endometriosis. (Ref. 4, p. 1080)

667. E. Ovarian cancer, probably in its initial stages, grows slowly and extensively and invades the stroma and germinal epithelium before involving the peritoneal cavity. Although the lymphatic channels in the ovary are involved at an early stage, lymph node metastases are not demonstrable until later in the course of the disease. Omental metastatic deposits are an early manifestation of dissemination of ovarian cancer, and may often be found before there is any evidence of a break on the ovarian surface. It is for this reason that omentectomy is considered an integral part of the operative treatment of this disease. (Ref. 4, p. 1143)

668. C. Laparoscopy is contraindicated in patients with serious cardiac or pulmonary disease, since anesthesia is contraindicated. Likewise, the procedure should not be undertaken in patients with intestinal obstruction, extensive herniations, abdominal scarring, or when it is not feasible to safely establish a pneumoperitoneum. (Ref. 5, p. 411)

669. A. Breast cancer is the most common cancer in women, at 76/100,000 women per year. About 1 of every 11 women can be expected to develop breast cancer. Approximately 90% of infiltrating breast cancers are first seen as infiltrating ductal or lobular carcinomas arising from carcinoma in situ lesions in the ductal system. Minimal lesions (less than 5 mm) can probably be cured because nodal metastasis is unlikely. At 1 cm, 20% have metastasized to nodes. (Ref. 4, p. 1156)

670. E. The physical findings of a loud cardiac murmur, edema, or rales require a thorough cardiac assessment of the patient. Readily overlooked, but equally significant, are the presence of vascular bruits and gallop rhythm, as they may be clues to significant cardiovascular disease. Patients with severe preoperative cardiac disease, as exemplified by electrocardiographic evidence of myocardial ischemia, heart block, or cardiac decompensation, may have significant postoperative electrocardiographic and/or clinical deterioration. Of cardiac patients, 12% show evidence of deterioration in their cardiac status after major surgery. (Ref. 5, p. 74)

671. A. Metabolic problems due to intravenous hyperalimentation are often related to the infusion of high concentrations of glucose. As a result of such an infusion, there is an increase in serum osmolarity, and oncotic pressure and dehydration of the extravascular and intracellular compartments. Osmotic diuresis results and, if uncontrolled, can result in hyperosmolar nonketotic acidoses and dehydration. Diuresis with renal loss of potassium can result in hypokalemia, which is often accentuated by the increased protein anabolism. Hypocalcemia is uncommon, but occasionally hypophosphatemia occurs. With the advent of the new amino acid solutions, hyperchloremia and hyperammonemia rarely occur. (Ref. 5, p. 121)

672. E. Of all the postoperative complications of pelvic surgery, one of the most sinister hazards is venous thrombosis and pulmonary embolism. Changes in blood coagulation in the postoperative patient, such as increased platelet adhesiveness and aggregation, as well as an increase in circulating platelets, favor the evolution of venous thrombosis. There is also an increase in such intrinsic coagulation factors as VIII, IX, and X within the first 96 hours of the surgical procedure. An increase in fibrinogen and circulating fibrinolysin inhibitors also occurs. Tissue trauma and necrosis liberate tissue thromboplastin, and this extrinsic factor contributes to venous thrombosis by accelerating the clotting mechanism. (Ref. 5, p. 104)

673. E. Mucinous cystadenomas of the ovary may attain an enormous size. They are multilocular or parvilocular and are composed of large, rounded, cystic compartments that are lined by columnar epithelium similar to that in the endocervix. The secretion of these tumors is an acid mucopolysaccharide. Mucinous cysts become secondarily malignant in about 10% of cases. The benign mucinous cystadenoma with papillary excrescences should be considered malignant until proven otherwise. (Ref. 5, p. 879)

674. A. Serous cystadenomas of the ovary are lined by cuboidal, often ciliated, epithelium, which is non-mucous-producing. These cysts may be multilocular, parvilocular, or unilocular. Papillary excrescences are common on the interior, but less common on the external, surface of the cyst wall. These papillary growths, when seen on the external surface of the cyst wall, can implant upon the serosal surface of the abdominal viscera, but do not usually produce ascites unless frank malignant change has occurred. These tumors are bilateral in about 15% of cases, and about 30% have the propensity for malignant transformation. (Ref. 5, p. 878)

675. E. Retention of urine may be due to the presence of some obstruction that is usually outside the urinary tract; to reflex spasm; to paralysis of the bladder, from overdistention; or to hysteria. Obstruction causing retention of urine is seldom urethral, as urethral stricture is rare in women. More often it is due to a pelvic tumor, the most common conditions being a retroverted gravid uterus, impacted fibroid tumor, pelvic hematocele, and hematocolpos. Reflex spasm of the sphincter mechanism may cause acute retention after vaginal operative procedures. A very sensitive urethral caruncle that gives rise to severe pain during micturition may also cause acute urinary retention. (Ref. 5, p. 102)

676. A. This clinical entity is most commonly encountered in infants, children, and postmenopausal women. In postmenopausal women, when not marked, it can readily be mistaken for a sessile urethral caruncle. A urethral prolapse consists of a partial or complete ring of urethral mucous membrane that has become everted through the urethral meatus. Its attachment to the margins of this aperture can be readily demonstrated, and the urethral orifice is centrally located. In elderly patients, urethral prolapse may produce symptoms of bladder irritability similar to those seen with urethral caruncles. In a small child, the edema and tissue reaction may produce urinary retention. (Ref. 5, p. 753)

677. E. Most children with imperforate hymen are brought to the gynecologist at the age of 13 to 15 years, when they develop molimina but no obvious menstrual flow. Lower abdominal pain and pelvic discomfort constitute the most common symptoms of this clinical entity. The most common urologic symptoms are dysuria, frequency, and urgency. Overflow incontinence and acute retention of urine may occur after cryptomenorrhea of several months' duration because of the pressure of the distended vagina upon the urethra. (Ref. 5, p. 743)

678. A. At the present time, the main value of colposcopy is the evaluation of patients with abnormal cytologic findings. It is possible to localize the lesion, evaluate its extent, and obtain a directed biopsy for histologic study. The false-negative rate of directed biopsy is about 0.3%, while the histopathology will show a more advanced lesion than indicated by colposcopy in about 3% of cases. The squamocolumnar junction may not be visualized in many postmenopausal women and about 15% of women under the age of 45 years. Colposcopy is ideal for the evaluation and management of gravid patients with abnormal cytologic findings. (Ref. 5, p. 768)

679. E. Prophylaxis is obviously of the utmost importance and consists of the early diagnosis and treatment of venous thrombosis. Of these patients who die with massive pulmonary embolization, 75% expire within one hour. A massive pulmonary embolus is usually preceded by asymptomatic nonlethal emboli, although some of these small emboli may cause pleuritic pain. Pulmonary infarction due to pulmonary embolism, although well known, is one of the least frequent complications of pulmonary embolism, and is present in fewer than one-third of the cases. Clinically, pulmonary infarction is associated with dyspnea, pleuritic chest pain, and hemoptysis or cough. Signs of pulmonary infarction include fever, tachycardia, tachypnea, and, less frequently, a friction rub. Leukocytosis, an elevated lactic dehydrogenase level, and the classic wedge-shaped density on the chest x-ray are frequently observed. (Ref. 5, p. 104)

680. B. Early squamous metaplasia is most often noted in puberty, early adolescence, and during the first pregnancy. It is considered that women who commence sexual activity at an early age, when the metaplastic process is most active, have a greater chance of developing cervical cancer. Studies have shown that women who have early intercourse and multiple sexual partners have a higher incidence of the disease. It is well established that promiscuous women, prostitutes, and divorced women have a higher incidence of cervical cancer than matched controls. In the Arab population, although circumcision of the male is a religious ritual, cervical cancer is as common as among non-Jewish women throughout the world. (Ref. 5, p. 760)

681. E. The application of hemodynamic investigative techniques to critically ill patients at the bedside have shed considerable light on the status of the circulation in early septic shock. Instead of the anticipated low cardiac output with systemic hypotension, nearly all patients studied have been found to have normal or elevated cardiac output with reduced peripheral vascular resistance. This is the clinical picture of warm shock, as the extremities are warm and dry owing to peripheral vasodilation. Systemic arterial hypoxemia is often present, but is often related to ventilation as arterial PCO_2 tensions are also reduced. Despite hypoxia, there is a narrowed arteriovenous oxygen difference, reflecting reduced oxygen consumption. The earliest clinical signs are those of respiratory distress with hyperventilation. Respiratory alkalosis is present, and because of hypoxia, metabolic acidosis develops with accumulation of lactate and pyruvate. (Ref. 5, p. 311)

682. E. Anorexia nervosa is manifested predominantly by weight loss, amenorrhea, and behavioral disturbances. Amenorrhea is present in virtually all female patients and may precede significant weight loss in 25%. Other features include bradycardia, hypotension, hypothermia, constipation, lanugo hair growth, hypercarotenemia, mild diabetes insipidus, and, in severe cases, dependent edema. (Ref. 3, p. 197)

683. A. Prostacyclin (PGI_2) is a prostaglandin and the lungs are a major source of this compound. It is a potent vasodilator and it inhibits platelet aggregation and adherence. By its action it keeps healthy blood vessels free of clots except when endothelial damage allows platelets to adhere. (Ref. 3, p. 354)

684. B. The majority of circulating estradiol and testosterone is bound to a specific β-globulin carrier, sex hormone binding globulin. The concentration of SHBG is decreased by corticoids, androgens, progestins, and growth hormone. It is increased by estrogens, hyperthyroidism, and pregnancy. (Ref. 3, p. 16)

685. B. Von Willebrand disease is a heterogenous group of functional disorders characterized by a mild or moderate increase in bleeding time and factor VIII deficiency, but no intrinsic platelet abnormality. Most variants are inherited as autosomal dominant disorders, but the worst is autosomal recessive. Factor IX deficiency is known as Christmas disease hemophilia B. (Ref. 1, p. 795)

686. A. Hematologic effects of oral steroidal contraceptive agents include an increased sedimentation rate, increased total iron-binding capacity due to the increase in globulins, and a decrease in prothrombin time. There is also a small increase in vitamin A and decreases in B vitamins, folate, and ascorbate. (Ref. 3, p. 479)

687. B. FSH stimulates estrogen production, which has an inhibitory effect on FSH secretion. At the level of the dominant follicle, FSH and estrogen stimulate granulosa cell production and FSH receptors, which increase androgenic aromatization. Estrogen levels rise sharply late in the follicular phase and feed back in a positive loop to increase LH. LH increases production of progesterone, androstenedione, and testosterone in the theca cells. The high midcycle androgen production causes atresia in the nondominant follicles, but is converted to estrogen in the dominant follicle. Under exposure to estrogen and FSH, the granulosa cells develop LH receptors late in the follicular phase and begin secreting progesterone. Sustained high levels of estrogen produce the dramatic rise in LH: the LH surge. The presence of progesterone and prostaglandins in follicular fluid with high LH and a resurgence of FSH increase proteolytic enzymes, which produce ovulation 34 to 36 hours after the LH surge. After ovulation, LH and FSH levels decline, but continued low levels of LH are necessary for the large progesterone production from the luteinized granulosa cells. (Ref. 3, p. 91)

688. E. Almost 80% of breast cancer presents initially as a mass usually (70%) found by the patient. The second most common sign is nipple discharge, which may be bloody, nonbloody, or any color. Less frequent initial signs include nipple retraction, skin edema, erythema, ulceration, pain, ecchymoses. An increasing number are detected by mammography in the absence of physical signs or symptoms. Radiographic findings include fine calcifications, stellate densities, and focal architectural changes in the parenchyma. (Ref. 4, p. 1156)

689. B. Adenomyosis is usually seen in multiparas in the fourth and fifth decades, while endometriosis is in younger nulligravidas who have delayed childbearing or are infertile. Both are associated with pelvic pain, worsening dysmenorrhea, and menorrhagia. Adenomyosis is suspected if the uterus is large, globular, and tender to palpation. Endometriosis is more likely if shoddy, tender nodules can be palpated on the uterosacral ligaments, or if hematochezia occurs with menses. (Ref. 5, p. 257)

690. A. Pelvic inflammatory disease is used to describe a spectrum of stages of pelvic infections caused by numerous organisms. Historically, *Neisseria gonorrhoeae* was the major isolate; but more recently, multiple aerobic and anaerobic organisms have been found. At present, the major offending organism appears to be *Chlamydia trachomatis*. This PID usually occurs during or near menses, starting as pelvic pain and tenderness. It produces high fever, leucocytosis with left shift, and elevated erythrocyte sedimentation rate. The pain tends to worsen and spread up the abdomen, even to the upper quadrants, to produce perihepatitis (Fitzhugh-Curtis syndrome) and abnormal liver function studies, but not altered pancreatic enzymes. The presence of the gonococcus can be confirmed by gram stain of the endocervix. (Ref. 5, p. 287)

691. E. Tubal pregnancy occurs in about 1/10,000 females and a little more than 1/100 live births. Early diagnosis improves the outcome and depends on a high index of suspicion. The classical triad of pain, bleeding, and an adnexal mass has been diagnostic in only 14% of cases. In the absence of nonclotting blood on culdocentesis, a quantitative β-HCG may be obtained serially. Early in pregnancy it should increase by 66% every 48 hours. If the serum level is 4000 mIU/mL, vaginal ultrasound should be performed; if it is 6000 or more mIU/mL, abdominal ultrasound can identify an intrauterine pregnancy. (Ref. 5, p. 429)

692. C. Infertility affects about 15% of American couples, and increases with age in women, reaching 70% by age 40, and is usually defined as an inability to achieve pregnancy in one year. In addition to a complete history and physical examination, the full evaluation entails assessing male factors, ovulation, cervical factor, uterine factor, tubal factor, and peritoneal factor. (Ref. 2, p. 263)

693. E. Amenorrhea may be due to a variety of etiologies. Careful history and physical exam should reveal the etiology in about 85% of cases. Confirmatory laboratory data can then be ordered specifically. Recent weight changes, especially loss with exercise, is associated with anovulation. In the presence of hyperestrogenic states, the endometrium should be evaluated for hyperplasia. Prolactinomas may be confirmed with elevated blood prolactin and CT scan. Uterine adhesions following curettage, especially associated with pregnancy, characterize Asherman's syndrome. (Ref. 2, p. 351)

694. E. Abnormal bleeding is one of the most frequently mismanaged problems encountered in gynecology. The goals of therapy are to stop the heavy, continuous bleeding and reestablish cyclic bleeding. History and physical examination will establish etiologies such as trauma, uterine fibroids, vaginal and cervical cancers, cervical polyps, and pregnancy-related bleeding. After ruling out pregnancy, the endometrium should be sampled prior to initiating therapy to stop the bleeding. It is appropriate to check for a coagulopathy and platelet function when checking the blood count. Ultrasound may be helpful in identifying submucous fibroids and perhaps polyps. Hysteroscopy and/or curettage may be necessary to make the diagnosis. Anovulatory bleeding can usually be controlled with cyclic progestins or oral contraceptives. If one of the goals is pregnancy, ovulation induction is appropriate. (Ref. 2, p. 378)

695. E. The initial laboratory evaluation of hirsutism should include serum testosterone, DHEAS, 17 OHP, prolactin, and test of thyroid production such as T4 or TSH. DHEAS levels less than 700 mg/dL and normal 17 hydroxyprogesterone assay rules out an adrenal problem. It should be noted that increased prolactin levels will increase DHEAS, increase SHBG, and therefore increase free testosterone levels. (Ref. 3, p. 241)

696. E. Late onset adrenal hyperplasia is diagnosed by the 17 hydroxyprogesterone assay. Late onset adrenal hyperplasia is the most common autosomal recessive disorder, surpassing cystic fibrosis and sickle cell anemia. Of women with hursitism, 1% to 5% will show a biochemical response consistent with this condition. This relative frequency dictates routine 17 hydroxyprogesterone screening. Less than 300 mg/dL rules out adrenal hyperplasia. (Ref. 3, pp. 244–248)

697. C. If total testosterone levels are greater than 200 mg/dL, an androgen-producing tumor must be suspected. Likewise, a history of rapidly progressive masculinization is suspicious for an androgen-producing tumor. Hirsutism associated with anovulation is slow to develop, usually covering a time period of at least several years. Androgen-producing tumors are vastly overrated and incredibly rare. (Ref. 3, pp. 251–252)

698. E. Almost all patients presenting with hirsutism represent excess androgen production caused by a steady state of anovulation. Treatment is directed at interrupting this steady state. (Ref. 3, pp. 252–257)

699. E. Spironolactone has multiple actions, inhibiting the ovarian and adrenal biosynthesis of androgens, competing for the androgen receptor in the air follicle, and directly inhibiting 5-α-reductase activity. 5-α-Reductase activity converts testosterone and androstenedione to dihydrotestosterone in the skin. (Ref. 3, p. 256)

700. A. Lack of Müllerian development (Mayer-Rokitansky-Kuster-Hauser syndrome) is the diagnosis for the individual with primary amenorrhea and no vagina. This is a relatively common cause of primary amenorrhea, more frequent than testicular feminization, and second only to gonadal dysgenesis. Approximately 33% have urinary tract anomalies, and 12% or more have skeletal anomalies, mostly involving the spine. (Ref. 3, pp. 182, 185)

References

1. Cunningham FG, MacDonald PC, Gant NF: *Williams Obstetrics,* ed 18. Norwalk, CT/San Mateo, CA, Appleton & Lange, 1989.

2. Jones HW, Wentz, AC, Burnett LS: *Novak's Textbook of Gynecology,* ed 11. Baltimore, Williams & Wilkins, 1988.

3. Speroff L, Glass RH, Kase NG: *Clinical Gynecologic Endocrinology and Infertility,* ed 4. Baltimore, Williams & Wilkins, 1989.

4. Danforth DN, Scott JR: *Obstetrics and Gynecology,* ed 5. Philadelphia, JB Lippincott Co, 1986.

5. Mattingly RF, Thompson JD: *Te Linde's Operative Gynecology,* ed 6. Philadelphia, JB Lippincott Co, 1985.